Farming System and Sustainable Agriculture

New Delhi-110 034

About the Author

Dr. R.K. Nanwal, Chief Agronomist (Retd.), Department of Agronomy CCS Haryana Agricultural University, Hisar-125004 (India). For being M.Sc (Ag.) in Agronomy from HAU, Hisar in 1981, he was awarded Ph.D. in Agronomy by IARI, N.Delhi in 1992. He did Post Doctorate specialized training in water resource management in Rainfed Agriculture at University of Georgia, USA. Attended many technical trainings, refresher courses and summer institutes in various aspects of agriculture for updating the knowledge. Eight such trainings were attended in India and one in USA. Acted as course coordinator/Director of various courses organized by counseling and placement centre/AAREM/Department of Agronomy etc. Designed 40 such courses. He has teaching experience of more than 35 years in the university. Developed new courses on Farming Systems, Sustainable Agriculture and Rainfed Agriculture. Guided 7 M.Sc and 8 Ph.D students as Major Advisor. He was member of advisory committee in number of post graduate students. He has published 9 books, 10 manuals, 7 book chapters, 153 full research papers, 75 general articles, 35 research notes and presented 60 papers in various national/international seminars/conferences. Received "Indian Council of Agricultural Research, N. Delhi Award" for outstanding Multidisciplinary team research in agriculture and allied sciences for the triennium 1994-96, awarded on July 16, 1997 by Union Minister of Agriculture, Govt of India. Awarded CCS Haryana Agricultural University "Best Teacher Award" (Indian Council of Agricultural Research sponsored) for the year 2005-2006. He was awarded "Rashtriya Gaurav Award" by India International Friendship Society on Nov.11, 2011. He was also awarded Best Citizens of India Award by International Publishing House, N. Delhi on Dec. 28, 2011.

Farming System and Sustainable Agriculture

R.K. Nanwal
Professor (Retd.)
Department of Agronomy
CCS Haryana Agricultural University
Hisar 125 004, Haryana, India

NEW INDIA PUBLISHING AGENCY
New Delhi-110 034

NEW INDIA PUBLISHING AGENCY

101, Vikas Surya Plaza, CU Block, LSC Market
Pitam Pura, New Delhi 110 034, India
Phone: + 91 (11) 27 34 17 17 Fax: + 91 (11) 27 34 16 16
Email: info@nipabooks.com
Web: www.nipabooks.com

Feedback at feedbacks@nipabooks.com

ISBN 978-93-95763-91-2

Composed and Designed by NIPA

Dedicated to my mother

SMT. SHANTI DEVI

Always a source of inspiration to me

Preface

The glory of Green Revolution proved to be a turning point in improving the food situation from "begging bowl to self-sufficiency". There has been a spectacular increase in food grain production of India from 50.8 million tons in 1950-51 to more than 275 million tons at present. Day-by-day, the agricultural scenario is changing fast and task of keeping pace with population growth is strenuous and challenging. Now days several emerging issues have been addressed to sustain the high productivity under different cropping and farming systems. The future of Indian agriculture depends on the development of appropriate farming system as applicable to resource poor farm families and as suited to different agro-ecological zones. Integrated farming system is not only a reliable way of obtaining fairly high productivity with substantial fertilizer economy, but also a concept of ecological soundness leading to sustainable agriculture.

A judicious mix of cropping systems with associated enterprises like dairy, poultry, piggery, fishery, sericulture, bee keeping, mushroom production etc suited to the given agro- climatic conditions and socio-economic status of the farmers would bring prosperity to the farmer. The future of Indian agriculture depends on the development of appropriate farming system as applicable to resource poor farm families and as suited to different agro-ecological zones. Integrated farming system is not only a reliable way of obtaining fairly high productivity with substantial fertilizer economy, but also a concept of ecological soundness leading to sustainable agriculture.

The book brings out in detail the agronomy course on farming systems and sustainable agriculture recommended for undergraduate students by 5th Dean's Committee and accepted by the ICAR for uniform adoption by all SAUs. The chapters included in this book are a mixture of theoretical and applied investigation. The book is comprehensive covering all the basics of farming systems. The author acknowledges his indebtedness to author of books and research articles from which most of the material in this introductory book has been drawn. In several cases, it has not been possible to obtain permission for reproduction for which the authors and publisher offer their sincere apologies. Inspite of the best efforts, it is possible that some errors may have crept into the compilation. The readers are requested to kindly let me know the mistakes so that these can be taken care of in the further edition.

I am thankful to my family especially my wife Mrs. Raj Rani for supporting me during the course of writing the book. I acknowledge with affection the love of my grandsons Shourya and Shlok which always keep me cheerful. Finally, I thank the publisher for bringing out this book so efficiently and promptly.

Author

Contents

1

Farming System - Concept, Scope Importance, Types and Factors Affecting Types of Farming Systems

The prosperity of any country depends upon the prosperity of its people. Food, clothing, shelter, health, education, security, roads, electricity and drinking water are the today's basic needs. Farmers produce food. About 70 % of Indian population is engaged in farming. Therefore, prosperity of India would depend upon the prosperity of its farmers. This in turn depends upon the adoption of improved technology and judicious allocation of resources (land, labour, capital, machinery etc). Human race depends more on farm products for their existence than anything else since food and clothing – the prime necessaries are products of farming. Even for industrial prosperity, farming forms the basic raw material. To sustain and satisfy as many as his needs, the farmers include crop production, livestock, poultry, fisheries, beekeeping etc. in their farms. Earlier subsistence was the important objective of farming. Farmers took many activities such as planting of fruit trees in their farms or on the common lands just for the welfare of mankind without expecting anything in return. The kind of farming where subsistence and welfare of mankind were the main objectives is designated as mixed farming. Farming today is also a mix of enterprises. However, higher profitability without altering ecological balance is important in farming. A set of agricultural activities organized while preserving land productivity, environmental quality and maintaining desirable level of biological diversity and ecological stability is designated as "Farming System". Here the emphasis is mainly on a system rather than on gross output.

In other words "farming system" is a resource management strategy to achieve economic and sustain agricultural production to meet diverse requirement of the farm household while preserving the resource base and maintaining high environmental quality. The farming system in its real sense will help to lift the economy of agriculture and standard of living of the farmers.

Farming system specially refers to a group/combination of enterprises in which the products and or the byproducts of one enterprise serve as the inputs for production of other enterprise. The waste of dairying like dung, urine, refuse etc. is used for preparation of FYM, which is an input in cropping systems. The straw obtained from the crops is used as fodder for cattle which in turn are used for different field operations for growing crops. Thus different enterprises of farming systems are highly interrelated.

Farming system takes into account the combination needs of the family- the economic factors like relative profitability of the technically feasible enterprises, availability of farm resources, infrastructure and institutions such as irrigation, marketing facilities including storage and transportation and credit besides the agro biological consideration namely interdependence, if any among various technically feasible enterprises and the performance of individual farmers.

Farming is defined as the way in which the farm resources are allocated to the needs and priorities of the farmers in his local circumstances which include-

1. Agro climatic condition such as the quantity, distribution and reliability of rainfall temperature, etc
2. Soil type, topography and
3. Economic and institutional circumstances like market opportunities, prices, institutional and infrastructure facilities and technology.

Definition

Different scientists have defined a farming system differently. However, many definitions, in general, convey the same meaning that it is strategy to achieve profitable and sustained agricultural production to meet the diversified needs of farming community through efficient use of farm resources without degrading the natural resource base and environmental quality. Relatively recent definitions include:

- Farming system is a resource management strategy to achieve economic and sustained agricultural production to meet diverse requirements of farm livelihood while preserving resource base and maintaining a high level of environment quality. Farming system is a set of agro economic activities that are interrelated and interact with themselves in a particular agrarian setting. It is a mix of farm enterprises to which farm families allocate its resources in order to efficiently utilize the existing enterprises for increasing the productivity and profitability of the farm. These farm enterprises are crop, livestock, aquaculture, agro forestry and agri-horticulture.
- Farming system is a mix of farm enterprises such as crop, livestock, aquaculture, agro forestry and fruit crops to which farm family allocates its resources in order to efficiently manage the existing environment for the

attainment of the family goal. Farming system represents an appropriate combination of farm enterprises (cropping systems horticulture, livestock, fishery, forestry, poultry) and the means available to the farmer to raise them for profitability. It interacts adequately with environment without dislocating the ecological and socio-economic balance on one hand and attempts to meet the national goals on the other.

- Farming system is a decision making unit comprising the farm household, cropping and livestock system that transform land, capital and labour into useful products that can be consumed or sold.

Key principles

- *Cyclic.* The farming system is essentially cyclic (organic resources – livestock – land – crops). Therefore, management decisions related to one component may affect the others.
- *Rational.* Using crop residues more rationally is an important route out of poverty. For resource-poor farmers, the correct management of crop residues, together with an optimal allocation of scarce resources, leads to sustainable production.
- *Ecologically sustainable.* Combining ecological sustainability and economic viability, the integrated livestock-farming system maintains and improves agricultural productivity while also reducing negative environmental impacts.

Scope of Farming System

Farming enterprises include crop, livestock, poultry, fish, sericulture etc. A combination of one or more enterprises with cropping when carefully chosen, planned and executed gives greater dividends than a single enterprise, especially for small and marginal farmers. Farm as a unit is to be considered and planned for effective integration of the enterprises to be combined with crop production activity.

Integration of Farm Enterprises depends on many factors such as:

1. Soil and climatic features of the selected area.
2. Availability of the resources, land, labor & capital.
3. Present level of utilization of resources.
4. Economics of proposed integrated farming system
5. Managerial skill of farmer.

The scope of IFS include pooling and sharing of resources/inputs, efficient use of family labour, conservation, preservation and utilization of farm biomass including non- conventional feed and fodder resources, effective use o f manure/ animal waste, regulation of soil fertility and health, income and employment generation for many people and increase economic resources. It improves space utilization and provides diversified products. The IFS is part of the strategy to ensure sustainable use of the natural resources for the benefit of present and future generations.

Although integrated farming is economically and environmentally sound, the motivation for integration will support the national policy of diversification of production. The following have been positively identified as advantages of an integrated farming system.

- Higher food production to equate the demand of the exploding population of our nation;
- Increased farm income through proper residue recycling and allied components;
- Sustainable soil fertility and productivity through organic waste recycling;
- Regular stable income through the products like egg, milk, mushroom, vegetables, honey from the linked activities in integrated farming;
- Inclusion of biogas to solve /mitigate the prognosticated energy crisis.

The opportunities provided by integrated farming:

Productivity: Includes productivity/ increased economic yield per unit area – per unit time by virtue of intensification of crops, agricultural crop rotation and allied enterprises.

Integrated Farming Systems provide opportunities as crop insurance cover as money round the year is obtained from different farm produces. It promotes technology Infusion Research and Development (R&D) integrated with indigenous/ Traditional knowledge.

Sustainability: In Integrated Farming Systems, organic supplementation through effective utilization of by-products of linked components as a measure is possible and this will certainly provide opportunity to promote soil health and to sustain the potentiality of the soil which is the production base.

Balanced food: In Integrated Farming Systems, components of different nature are linked enabling production of different sources of nutrition, namely, protein, carbohydrates, fats, minerals, vitamins, etc from the same unit. It provides opportunity to mitigate malnutrition problem of the farmers.

Pollution abatement: In crop based activity, some of the organics are left as waste materials which in turn pollute the environment on decomposition. Application of huge quantities of fertilisers, pesticides, weedicides, insecticides, etc. pollutes soil, water and air. Much of the wastes could be converted/recycled to some other forms of economic/ecological/social value, under the Integrated Farming System.

Agro-industries: Integrated farming also provides opportunities for agri-oriented industries, tourism and related tourism based activities.

The other benefits of farming systems are:

- Promotion of Agro-Industry.
- Increased Input Efficiency.
- Cost Minimization for Input Use.
- Increased Employment.
- Fodder Security for Livestock.
- Recycling.
- Continuous Income Round the Year.
- Energy Saving.

Farming System Concept

Farming system is a complex inter related matrix of soil, plants, animals, implements, power, labour, capital and other inputs controlled in parts by farming families and influenced to varying degree by political, economic, institutional and socials forces that operate at many levels. Thus farming system is the result of a complex interaction among a number of interdependent components. Farm activities interact with market forces (socio-economic) and ecosystem (biophysical) for purchasing inputs and disposing outputs by utilizing and degrading natural resources (land, water, air, sunshine etc.).

A farm is a system in that it has **inputs, processes** and **outputs**

Inputs - these are things that go into the farm and may be split into *Physical Inputs* (e.g. amount of rain, soil) and *Human Inputs* (e.g. labour, money etc.)

Processes - these are things which take place on the farm in order to convert the inputs to outputs (e.g. sowing, weeding, harvesting etc.)

Outputs - these are the products from the farm (i.e. wheat, barley, fruits) Depending on the type of farming e.g. arable/pastoral, commercial/subsistence, the type and amount of inputs, processes and outputs will vary. You need to make sure you are able to define and give examples of Inputs, Processes and Outputs in farming systems.

Income through arable farming alone is insufficient for bulk of the marginal farmers. The other activities such as dairying, poultry, sericulture, apiculture, fisheries etc. assume critical importance in supplementing their farm income.

The modern agriculture emphasize too more dimensions *viz*., time and space concept. Time concept relates to increasing crop intensification in situation where there is no constraint for inputs. In rainfed areas where there is no possibility of increasing the intensity of cropping, the other modern concept (space concept) can be applied. In space concept, crops are arranged in tier system combining two or more crops with varying field duration as intercrops by suitably modifying the planting method. Income through arable cropping alone is insufficient for bulk of the marginal farmers. Activities such as dairy, poultry, fish culture, sericulture, bio-gas production, edible mushroom cultivation, agro-forestry and agri-horticulture, etc., assumes critical importance in supplementing farm income. It should fit well with farm level infrastructure and ensures full utilization of bye-products. Integrated farming system is only the answer to the problem of increasing food production for increasing income and for improving the nutrition of small scale farmers with limited resources.

Livestock raising along with crop production is the traditional mixture of activities of the farmer all over the country, only the nature and extent varies from region to region. It fits well with farm level infrastructure, small land base and abundant labour of small man and ensures full utilization of by products.

Sustainability is the objective utilization of inputs without impairing the quality of environment with which it interacts. Therefore, it is clear that farming system is a process in which sustainability of production is the objective.

The overall objective is to evolve technically feasible and economically viable farming system models by integrating cropping with allied enterprises for irrigated, rained, hilly and coastal areas with a view to generate income and employment from the farm.

The specific objectives are:

1. To identify existing farming systems in specific areas and access their relative viability.
2. To formulate farming system model involving main and allied enterprises for different farming situations.
3. To ensure optional utilization and conservation of available resources and effective recycling of farm residues within system and
4. To maintain sustainable production system without damaging resources/ environment.

Goals of IFS

The goals of integrated farming system (IFS) is to through rejunvenation of the system's productivity and ecological equilibrium through the reduction in the build up of pest and diseases through natural cropping system management and the reduction in the use of chemicals (in- organic fertilizers and pesticides).

Types and systems of farming system and factors affecting types of farming

Many farms have a general similarity in size, products sold and methods followed is called a type of farming or when farms are quite similar in kind and production of the crops and live stock that are produced and methods and practices used in production, the group is called as type of farming. On the basis of the share of gross income received from different sources and comparative advantage, the farming systems may be classified as follows:

A) According to the Size of the Farm

a) Collective farming

b) Cultivation farming: (i) small scale farming (ii) large scale farming.

B) According to the Proportion of Land, Labour and Capital Investment

a) Intensive cultivation.

b) Extensive cultivation.

C) According to the Value of Products or Income or on the basis of Comparative Advantages

i) Specialized farming.

ii) Diversified farming.

iii) Mixed farming.

iv) Ranching.

v) Dry farming.

D) According to the Water Supply

i) Rainfed farming.

ii) Irrigated farming.

E) According to

i) Type of Rotation

a) Lay system (i) unregulated lay farming (ii) regulated lay system.

b) Field system.

c) Perennial crop system.

ii) Intensity of the Rotation

a) Shifting cultivation.

b) Lay or fallow farming.

c) Permanent cultivation.

d) Multiple cropping.

F) Classification According to Degree of Commercialization

a) Commercialized farming.

b) Partly commercialized farming.

c) Subsistence farming.

G) Classification According to Degree of Nomadic

a) Total nomadic

b) Semi nomadic

c) Partial nomadic

d) Transhumant

e) Stationary animal husbandry

H) Classification According to Cropping and Animal Activities

a) Monocropping with livestock

b) Double cropping with livestock

I) Classification According to Implements Used for Cultivation

a) Spade farming

b) Hoe farming

c) Mechanized or tractor farming

Classification of Farming Systems- According to size of farm

According to the size of the farm:

1. Collective Farming

It includes the direct collection of plant products from non-arable lands. It may include either regular or irregular harvesting of uncultivated plants; honeying and fishing usually go hand in hand with collection. Actual cultivation is not needed. The natural products like honey, gum, flower etc are collected. Such plant product may be collected from forestry area.

2. Cultivation Farming

In this system, farming community cultivates the land for growing crops for obtaining maximum production per unit area.

i) Small Scale Farming

In this type, the farming is done on small size of holding and other factors of production are small in quantity and scale of production is also small.

Advantages

i) Intensive cultivation is possible

ii) Labour problem do not affect the production.

iii) It is easy to manage the farm

iv) There is less loss due to natural calamities like frost, heavy rainfall, and diseases

v) Per unit output increases

Disadvantages

i) Small- scale farming cannot take advantages of various economic measures:

ii) Cost of production per unit is more

iii) Mechanization of agriculture is not possible.

iv) Farmer does not get employment round the year.

ii) Large scale farming

When farming is done on large size holding with large amount of capital, large labour force, large organization and large risk are called large- scale farming.

In other words when the factors of production are large in quantity, the small of farming is said to be large. In India 40 to 50 hectares land holding may be said large scale farming but in countries like America, Canada and Australia even 100 ha. Farms are also called as small farms.

Advantages

1. Production of large scale farming is more economical. The cost of production per unit is less.
2. Per unit production is increased.
3. Better marketing of agricultural products is possible. Processing, transportation, storage, packaging of produce is economical.
4. Costly machine like tractor, combined harvester can be maintained on the farm.
5. Subsidiary occupation such as dairy, poultry, bee keeping based on maintained on the farm.
6. Proper utilization of factors of production is possible.
7. Research work is possible.
8. It increases bargaining power of people.

Disadvantages

1. If demand of produce decreases and production exceeds the market demand there will be more loss to the large farm.
2. In case of labour strike there will be more loss on the farm.
3. Due to natural calamities like frost, drought, flood, insects and diseases the large farm will suffer a lot.
4. It will be difficult to manage large scale farm.

Classification of Farming Systems- According to Production of Land, Labour and Capital investment:

The farmer on given plot of land obtains more or less definite quantity of yield of any particular time.

If he wants to increases his output he can either

i) Bring more land under cultivation

ii) Apply more labour and capital to the same piece of land.

a) Intensive Cultivation

In intensive cultivation more labour and capital is used in the same piece of land. In other words land remains fixed in quantity while other factors are increased. If the same land is rare due to population pressure, while labour and capital are comparatively cheap, intensive cultivation is preferred than extensive cultivation. The application of intensive cultivation method depends mainly upon-

i) Increasing population and

ii) Technical improvement.

In the earlier stages of development population was small and technical knowledge of agriculture was also limited hence extensive method was adopted but, as population increases intensive cultivation become necessary and improvement in technique make its adoption is possible.

b) Extensive Cultivation

When more area is brought under cultivation to increases the output it is termed as extensive cultivation. In extensive cultivation land is chiefly available but availability of other factors increases less proportionately. A cultivator wishing to increases his output may follow either intensive method or extensive method but the selection of these two methods is based on cost. If following extensive cultivation than by following intensive cultivation can raise the additional output more cheaply, extensive method of cultivation will be useful. If on the other hand intensive cultivation seems to be the cheaper method he will naturally adopt it. If land is cheaper and it can be had at a normal cost while labour and capital are comparatively costlier, extensive cultivation will be cheaper method of obtaining increased output. In early times when land was plentiful extensive cultivation was followed. The extensive and intensive cultivation go side by side in a country for a certain period of time and afterwards intensive cultivation may become more important method. In most of the countries extensive and intensive methods of cultivation generally go hand in hand.

According to the Value of Products or Income or on the basis of Comparative Advantages

Specialized Farming

The farm in which 50% or more income of total crop production is derived from a single crop is called specialized farming or The farm in which only single crop is cultivated for selling in the market and the income of the farm depends mainly on that crop is called specialized farming by Hopkins.

According to the definition if 50% income is derived from paddy from any farm this is called paddy farm similarly sugarcane farm wheat farm, vegetable farm, orchard farm etc.

Advantages and Disadvantages of Specialized Farming

Advantages

1. Better use of land: More profitable to grow crops on land best suited to it. e.g. jute growing or cultivation on swampy land in west Bengal.
2. Better marketing: it allows grading, processing, storing, transporting and financing the produce.
3. Less equipment and labour.
4. Costly and efficient machinery can be kept: A wheat harvester thresher can be maintained in a highly specialized wheat farm.
5. The efficiency and skill of the labour increased: Specialization allows a man to be more efficient and expert at doing a few things.
6. Farm records can be maintained easily.
7. Intensity of production leads to relatively large amount of output.
8. Better management: fewer enterprises on the farm are liable to be less neglected and sources of wastage can easily be detected.

Disadvantage

These disadvantages of specialization are evident when the farmer realizes that "all his eggs are in one basket".

1. There is greater risk: When failure of crop and decreasing market price of the product, demand in market of product.
2. It is not possible to maintain soil fertility-lack of crop rotation.
3. The productive resources i.e. land; labour and capital are not fully utilized.
4. Irregular income of the farm as they get income only once or twice in a year.
5. Proper Utilization of resources is not possible.
6. By product of crop are not property utilized, as numbers of livestock's are less in number.
7. Due to specialization of a single enterprise, the knowledge about other enterprises vainness.
8. Does not help in supplying all the food needs of the family members of the farmer.

Diversified Farming and Its Advantages

Diversified Farming

A diversified farm is one that has several production enterprises or sources of income but no source of income equal as much as 50% of the total income from that source on such farm farmers depends on several sources of incomes. It is also called as general farming.

Advantages

1. Better use of land, labour and capital: Better area land through adoption of crop rotations, steady employment of farm and family labour and more profitable use of equipment are obtained in diversified farming.
2. The farmer and labour engaged all the year round in different activities.
3. Less risk to crop failure and market price of the product.
4. The byproducts of this farm can utilize properly as cattle, poultry, birds, etc. are reared with crop production.
5. Regular and quicker return is obtained from various enterprises.
6. Soil erosion can be checked as land kept under cultivated throughout the year
7. Soil fertility can be checked as land kept under cultivated throughout the year.
8. Diversified farming is less risky than specialized farming.
9. Best use of all equipments.

Disadvantages

1. Do not fetch desirable profit so long as co-operative marketing facility is not there.
2. Proper inspection of different enterprises is difficult.
3. It is not possible to farmer to maintain all types of machinery required for different crops.
4. The wastage of farm in any farm is difficult to detect.

Mixed Farming and Its Advantages

Mixed farming: Mixed farming is one which crop production is combined with the rearing of livestock. The live stock enterprises are complementary to crop production; so as to provide a balance and productive system of farming. In mixed farming at least 10% of its gross income must be contributed by livestock activity. The upper limit being 45% under Indian condition. So the farm on which at least 10 to 49% income is found from livestock is called mixed farm. In mixed farming cow and buffaloes are included with crop production. If farmers are rearing cows, buffaloes, sheep goat, and fisheries with crop cultivation this type of farming is called diversified farming.

The scope of mixed farming to combination of crops and their complementary livestock enterprises of mixed farming would certainly include a vast majority of our farms, establishing a complementary relationship between crop and livestock enterprises.

Enterprises	Contribution to gross income of farm	Farming type
1. Cow and Buffalo only	10 to 49%	Mixed farming
2. Cow, buffalo and poultry	10 to 49%	Diversified farming

Advantages

1. It offers highest return on farm business, as the by products of farm are properly utilized.
2. It provides work throughout year.
3. Efficient utilization of land, labour, equipment and other resources.
4. The crop by products such as straw, bus, fodder etc. is used for feeding of livestock and in return they provide milk.
5. Manures available from livestock maintain soil fertility.
6. It helps in supplying all the food needs of the family members.
7. Intensive cultivation is possible.

8. If one source of income is lost he can maintain his family from other source of income.
9. Milk cattle's provide draft animals for crop production and rural transport.
10. Mixed farming increases social status of the farmer.

In India the livestock is much closed connected with agriculture because animal power is the main source of power in agriculture. FYM is the main source for maintaining soil fertility and animals make good use of subsidiary and by-products on farms and in turn they provide milk under such circumstances mixed farming will most suit in Indian conditions.

Disadvantages

1. Indigenous method of cultivation is used till now.
2. Draft and milch animals should be sold when they fail in production.
3. Healthy calf should be reared to replace age old animals.

Requirement of Mixed Farming

i) Complicated management practices.
ii) Sound cropping scheme.
iii) Good cattle in suitable number.
iv) Transport facility.
v) Marketing facilities.

Classification of Farming System-According to Water Supply

According to the Water Supply:
i) Rainfed farming
ii) Irrigated farming

i) Rainfed Farming

Agriculture mainly depends on the rainfall in most part of the country. 80% of the total cultivated arable land is rainfed. Rainfed farming is very risky system of farming where the success of the crop depends on the cycle of the monsoon. Timely rainfall is the pre-requisite of this farming. The uneven rainfall is quite detrimental to crop production.

Characteristics of Rainfed Farming

1. Crop and varieties, which can withstand moisture stress should be cultivated.
2. Kharif crops sown after receiving monsoon.

3. Not possible to adopt improved methods of cultivation only one or two crops are grown.
4. Crops rotation is not followed.
5. Soils of these areas are deficits in nutrient.
6. Mixed cropping should be practiced and adopt deep-rooted crops.
7. Short duration varieties fit well in rain fed areas.
8. The crops that are tolerant to drought should be cultivated.
9. Soil moisture should be preserve by mulching, FYM application.
10. Soil erosion which may be called "Creeping death" of the soil is a worldwide problem, so necessary measures should adopt to keep the soil productive.

Principles of relevant components of environmentally sustainable farming systems should include.

1. Reduce soil erosion and improving soil conservations.
2. Inclusion of legumes and cover crops in crop rotations.
3. Agro-forestry as an alternate land use system and
4. Judicious use of organic waste.

ii) Irrigated Farming

The crop can be grown throughout the year; moisture is not a limited factor.

Characteristics

1. Round the year cropping pattern becomes possible.
2. Intensive cropping is possible.
3. Production can be increased by proper utilization of productive resources.
4. Crop rotation can be executed properly due to adequate irrigation facility.
5. Manuring is safely done in irrigated crop.
6. The field experiment is possible, because of timely irrigation facility.

Classification According to Type of Rotation

According to Type of rotations:

a) Lay System:
   i) Unregulated laid farming.
   ii) Regulated lay system.
b) Field System.
c) Perennial Crop System.

I) Type of Rotation

The world rotation has two meanings according to the time period involved. There is long term alteration between various retypes of land use such as arable farming, tree farming, grassland use etc. in this rotation means the sequence of this basic type of land use on a given field. Within arable farming there is also the term crop rotation which means the short- term sequences of different arable crops on one field.

a) Lay System

In this system, several years of arable farming are followed by several years of grassed and legumes utilized for livestock production.

i) **Unregulated Lay Farming** : In this system natural vegetation grasses, bushy growth on pasture is allowed to grow during the period of fallow. This is an improved managed pasture.

ii) **Regulated Lay System:** During the period of fallow, certain types of grasses are grown or planted. These are the well managed pasture with fencing and adopting rotational grazing system.

b) Perennial Crop System

The crop which covers the land for many years e.g. Tea, Coffee, sugarcane. In some cases tree crops (oil palm, rubber) are alternated with fallow in other with arable farming, grazing etc.

II) Intensity of the Rotation

a) Shifting cultivation
b) Lay or fallow farming
c) Permanent cultivation
d) Multiple cropping

II) Intensity of the Rotation

It is denoted by" R" simple and appropriate criteria for classification, which gives the true relationship between crop cultivation and following within the total length of one cycle of utilization.

$$R = \frac{\text{No.of years of cultivation}}{\text{Length of the cycle of land utilization}} \times 100$$

The length of the cycle is the sum of the number of years of arable in farming the number of fallow years "R" indicates the production of the area under cultivation in relation to a total area available for arable farming.

a) Shifting cultivation: R< 33%
In the case there is more number of years of fallow than actual cultivation.

b) Lay or fallow farming: In this case R< 66% and 33%

c) Permanent cultivation; in this large area is cultivated and small area is left fallow. R> 66%.

d) Multiple cropping: where R = > 100%. If R= 150% means 50% area is under two crops in a year. If R = 300% means three crops in a year are being grown.

Classification According to Degree of Commercialization

Depending Upon the Produce Sold in the Market for Earning Money:

a) Commercialized farming.
b) Partly commercialized farming &
c) Subsistence farming.

a) **Commercialized Farming:** More than 50% of the produce is for sale.

b) **Partly Commercialized Farming:** More than 50% of the value of produce is for home consumption.

c) **Subsistence Farming**
Virtually there is no sale of crop and animal products, but used for home consumption. Subsistence farming is a type of farming where the farmers of our country cultivate the crop in their land for the livings. Hence, the holding is small in size; so improved method of cultivation is not possible. They fail to meet the total requirement. They reared cattle, poultry, along with crop cultivation in limited land meet their requirement.

Advantages

1. Utilizing productive resources profitably.
2. Farmers with their family members engaged though the year as they rearing cattle, poultry etc.
3. Farmer meet their demand from the income from cattle, poultry etc.
4. By product used properly.

Disadvantages

1. Fails to adopt improved crop cultivation technique do to small holding.
2. Cultivation mainly depends on monsoon rain.
3. Procurement of seed, fertilizer as and when required is difficult.
4. Income of this farm is very low.

2

Farming System Components and Their Maintenance

Various components of farming systems are as under

1. Cropping system

A system is set of elements which depend on each other and interacting among themselves. Farming system consists of several enterprises like cropping system, dairying, piggery, poultry, fishery, bee keeping, etc. These enterprises are interrelated. The end products and wastes of one enterprise are used as inputs in other. The wastes of cow, dung used as FYM in crop production and straw is used as feeding material for cow. Bullock is used for field preparation.

Cropping system is an important component of a farming system. It represents cropping pattern used on a farm and their interaction with farm resources, other farm enterprises and available technology which determine their makeup.

Cropping pattern: means the proportion of area under various crops at a point of time in a unit area. It indicates the yearly sequence and spatial arrangement of crops and fallow in an area. It indicates the early sequence and spatial arrangement of crops and fallow in an area. Crop sequence and crop rotation are generally used synonymously. Crop rotation refers to recurrent succession of crops on the same piece of land either in a year or over a longer period of time. Component crops are so chosen so that soil health in not impaired.

Types of cropping system

Monocropping

Monocropping refers to growing of only one crop on a piece of land year after year. e.g. under rain fed conditions sorghum is grown year after year.

Multiple Cropping

Growing two or more crops on the same piece of land in one calendar year is known as multiple cropping. It is intensification of cropping in space and time dimensions. It includes intercropping, mixed cropping and sequence cropping.

Intercropping

Intercropping is growing two or more crops simultaneously on the same piece of land with a definite row pattern. For example growing maize+ green gram in 2:1 ratio

Mixed cropping: is growing two or more crops simultaneously intermingled without any row pattern. It is common practice in most of dry land areas

Sequence cropping: sequence cropping can be defined as growing of two or more crops in sequence on the same piece of land in a farming year.

Parallel cropping

Cultivation of such crops which have different natural habit and zero competition e.g Black gram/green gram+maize. The peak nutrient demand period for green gram is around 30-35 DAS while it is 50 DAS for maize.

Multi-storied / multi tiered / multilevel cropping

Cultivation of two or more than two crops of different heights simultaneously on a certain piece of land in a certain period e.g. sugarcane+ mustard + onion.

Efficient Cropping System

Efficient cropping system for a particular farm depends on farm resources, farm enterprise and farm technology. The farm resources include land, labor, water, capital, and infrastructure. When land is limited, intensive cropping is adopted to fully utilize available water and labor. When sufficient and cheap labor is available, vegetable crops also include in the cropping system as they require more labor. Capital intensive crops like sugarcane, banana, turmeric etc... Find a place in the cropping system when capital is not a constraint. In low rainfall regions (<750mm/ annum) monocropping is followed and when rainfall is more than 750 mm intercropping is practiced. With sufficient irrigation water triple and quadruple cropping is adopted when other climatic factors are not limiting farm enterprises like dairying, poultry etc., also influence the type of cropping system. When the farm enterprise includes dairy, the cropping system should contain fodder cops as components. Change in cropping system takes place with the development of technology. The feasibility of growing four crop sequence in gangetic alluvial plains gave impetus to multiple cropping.

What are the benefits of cropping systems?

1. Maintain and enhance soil fertility: Some crops are soil exhausting while others help restore soil fertility. However, a diversity of crops will maintain soil fertility and keep production level high.

2. Enhanced crop growth: Crops may provide mutual benefits to each other for. for example, reducing lodging, improving winter survival, or even acting as windbreaks to improve growth.
3. Minimize spread of disease: The more diverse the species of plants and the longer period before the soil is reseeded with the same crop, the more likely disease problems will be avoided.
4. Control weeds: Crops planted at different times of the year have different weed species associated with them. Rotating crops helps prevent build up of any one serious weed species. The more different growth cycles the crops have in your rotation, the fewer weeds will be able to adapt to the field conditions.
5. Inhibit pest and insect growth: Changing crops each year to unrelated species can dramatically reduce the population of pests and insects. Corp, crop rotation frequently eliminates their food source and changes the habitat available to them.
6. Increase soil cover: Growing a diversity of crops helps keep field sizes smaller, which increases soil cover, improves solar radiation capture and reduces erosion.
7. Use resources more efficiently: Having a diverse group of crops helps to more efficiently use the available resources, natural resources, such as nutrients, sunlight and water in the soil, are evenly shared by plants over the growing period, minimizing the risk for nutrient deficiencies and drought. Other resources, such as labor, animal draft power, and machinery are also utilized more efficiently as the time and effort spent in planting and harvesting crops are more spread out over the harvesting period.
8. Reduce risk for crop failure. Having a diverse group of crops helps prevent total crops equally. It also reduces food security concerns, as well as the amount of money required to finance production.
9. Improved food and financial security. Choosing an appropriate and diverse number of crops will lead to a more regular food production throughout the year. With a lower risk for crop failure, there is a greater reliability on food production and income generation.

2. Dairy

Cattle rearing in India are carried out under a variety of climatic and environmental conditions. The cattle are broadly classified into four groups:

Draft breeds: The bullocks of these breeds are good draft animals, but the cows are poor milkers. e.g. Nagore, Hallikar, Kangeyan, Malı.

Dairy breeds: The cows are high milk yielders but the bullocks are of poor draft quality e.g., Sahiwal, Sindhi, Gir.

Dual purpose: The cows are fairly good milkers and the bullocks are with good draft work capacity e.g. Haryana, Ongole and Kankrej.

Exotic breeds: The exotic breeds are high milk yielders. e.g. Jersey, Holstein Friesian, Aryshire, Brown Swiss and Guernsey.

Buffaloes: Important dairy breeds of buffalo are Murrah, Nili Ravi (which has its home tract in Pakistan), Mehsana, Suti, Zafrabadi, Godavari and Bhadawari. Of these Godavari has been evolved through crossing local buffaloes in coastal regions of AP with Murrah.

Housing: Each cow requires 12-18 sq m space and the buffaloes need 12-15 sq m. it is important to provide good ventilation and an open shed of housing is always preferable. Dairy building should be located at an elevated place to facilitate easy drainage. The floor should be rough and have gradient of 2.5 cm for every 25 cm length.

Breeding and maintenance: The cows remain in milk for 9-10 months, the average calving interval being 16-18 months. A cow does not require more than 6-8 weeks of dry period. From the economic point of view, cow should ordinarily be bred during the second and third months after calving. In weak animals and heavy milkers, breeding may be delayed by 1 or 2 months. Cattle come to heat in more or less regular cycles of about 21 days which lasts for about a day. The best time to serve a cow is during the last stage of heat. If artificially inseminated, a second insemination may be given within 6-8 hours after the first insemination. The gestation period varies with individual cows and normally it is about 280 days.

In the case of buffaloes, the lactation period last for 7-9 months. She buffaloes come to heat every 21- 23 days. The gestation period is 310 days. Calf rearing is very important in the case of buffalo maintenance. Since they require abundant water, wallowing is required. Regular de-worming is needed for buffalo maintenance.

Under Indian conditions, cattle commonly mature at the age of about three years. This period can, however, be reduced by six months under well managed herd.

Feeding: Cattle feed generally contains fibrous, coarse, low nutrient straw material called roughage and concentrates.

Roughages: Dairy cattle are efficient use of roughages and convert large quantities of relatively inexpensive roughage into milk. Roughages are basic for cattle ration and include legumes, non- legume hays, straw and silage of legume and grasses.

Concentrates: grains and byproducts of grains and oilseeds constitute the concentrates. They are extensively used in dairy cattle ration. These include cereals (maize, sorghum, oats, barley), cotton seeds, industrial wastes (bran of wheat, rice, gram husk) and cakes of oilseeds (groundnut, sesame, rapeseed, soybean, linseed)

Vitamins and mineral mixtures: It is advisable to feed a supplement containing vitamins A and D. Minerals mixtures containing salt, Ca and P should also be provided in the ration.

The ration per animal per day normally includes concentrates at 1 kg for 2 litres of milk yield, green fodder at 20-30 kg, straw 5-7 kg and water 32 litres.

3. Poultry

Poultry is one of the fastest growing food industries in the world. Poultry meat accounts for about 27% of the total meat consumed worldwide and its consumption is growing at an average of 5% annually. Poultry industry in India is relatively a new agricultural industry. Till 1950, it was considered a backyard profession in India.

Breeds: Specific poultry stocks for egg and broilers production are available. A majority of the stocks used for egg production are crosses involving the strains or inbred lines of white Leghorn. To a limited extent, other breeds like Rhode Island Red, California Grey and Australop are used. Heavy breeds such as white Plymoth Rock, White Cornish and New Hampshire are used for cross-bred broiler chickens. Hence, it is essential to consider the strain within the breed at the time of purchase. Several commercial poultry breeders are selling day old chicks in India. It is best to start with the day old chicks.

Housing: Adequate space should be provided for the birds. Floor area of about 0.2 m^2 per adult bird is adequate for light breeds such as white Leghorn. About 0.3-0.4 m^2 per bird is required for heavy breeds. The house should have good ventilation and reasonably cool in summer and warm during winter, it should be located on well-drained ground from flood waters.

Feed: The feed conversion efficiency of the bird is far superior to other animals. About 60-70% of the total expenditure on poultry farming is spent on the poultry feed. Hence, use of cheap and efficient ration will give maximum profit. Ration should be balanced containing carbohydrates, fats, minerals and vitamins. Some of the common feed stuffs used for making poultry ration in India are:

Cereals: Maize, barley, oats, wheat, pearl millet, sorghum, rice-broken. Cakes/ meal: oil cakes, maize-gluten-meal, fish meal, meat meal, blood meal. Minerals/ salt: Limestone, Oyster shell, salt, manganese.

From the day old to 4 weeks of age, birds are fed on starter ration and thereafter finisher ration, which contain more energy and 18-20% protein. Feed may be given 2-3 times. In addition to the stuffs, antibiotics and drugs may also be added to the poultry ration. Laying hens are provided with oyster shell or ground limestone. Riboflavin is particularly needed.

Maintenance: the chicks must be vaccinated against Ranikhet diseases with F1 Strain vaccine within the first 6-7 days of age. One drop of vaccine may be administered in the eye and nostril. When chicks get the optimum body weight of 1.0-1.5 around six weeks, they can be marketed for broiler. Hens may be retained for one year for production i.e. upto the age of 1 to 2 years. After that they are disposed off for table purpose. IT may not be economical to keep the hens beyond 2 years since egg production would get reduced. One hen is capable of laying 180-230 eggs in a year starting from the sixth month. In addition, a laying hen produced about 230 g of fresh droppings (75% moisture) daily.

4. Bee keeping

Bee keeping is one of the most important agro-based industries which does not require raw material from the artisan like other industries. Nectar and pollen from flowers are the raw materials which are available in plenty in nature. Bee keeping can even be started with a single colony.

Species: there are two species, *Apis cerana* and *A. mellifera* are complementary to each other but have different adaptations. *A cerana* is better acclimatized to higher altitudes of the Himalayan region. *A. mellifera* is more profitable in the plains.

Management: The bee keeper should be familiar with the source of nectar and pollen within his locality. The most important source of nectar and pollen are maize, great millet, bulrush, sunflower and palm. The beginner should start with 2 and not more than 5 colonies. A minimum of two colonies is recommended because in the event of some mishap, such as the loss of the queen occurring in one, advantage may be taken with the other.

The hive consists of floor board, brood chamber, top cover, frames and entrance rod. These parts can easily be separated. The hive may be double walled or single walled. The single walled hive is light and cheap.

The most suitable time for commencing bee keeping in a locality is the arrival of the swarming season. Swarming is a natural tendency of bees to divide their colonies under conditions that are generally favourable for the survival of both parent colony and the swarm. This occurs during the late spring or early summer.

Honey collection: Honey should have good quality to meet the national and international standards. Qualities such as aroma, colour, consistency and floral sources are important. Proper honey straining and processing are needed to improve the quality of the produce. Honey varies in the proportion of its constituents owing to the differences in the nectar produced by different plants. The nectar collected by bees is processed and placed in comb cells for ripening. During the ripening, sucrose is converted into glucose and fructose by an enzyme called invertase which is added to it by the bees. Honey is an excellent energy food with an average of about 3500 calories per kg. It is directly absorbed into the human blood stream, requiring no digestion.

5. Aquaculture

Ponds serve various useful purposes, viz. domestic requirement of water, supplementary irrigation source to adjoining crop fields and pisciculture. With the traditional management, farmers obtain hardly 300-400 kg of wild and culture fish per ha annually. However, poly-fish culture with the stocking density of 7500 fingerlings and supplementary feeding will boost the total biomass production.

Pond: the depth of pond should be 1.5 to 2.0 m. this depth will help for effective photosynthesis and temperature maintenance for the growth of zoo and phytoplanktons. Clay soils have high water retention capacity and hence are best suited for fish rearing. Pond water should have appropriate proportion of nutrients, phosphate (0.2 -0.4 ppm), nitrate (0.06-0.1 ppm) and dissolved oxygen (5.0-7.0). Water should be slightly alkaline (pH 7.5-8.5). if the pH is less than 6.5, it can be adjusted with the addition of lime at an interval of 2-3 days. Higher pH (>8.5) can be reduced with addition of gypsum. Application of fresh dung may also reduce high pH in the water.

Soil of the pond should be tested for N and P content. If the nutrient content is less, nitrogenous fertilizers like ammonium sulphate and urea and phosphate fertilizers like super phosphate can be added. Organic manures such as FYM and poultry droppings may also be applied to promote the growth of phyto and zooplanktons.

Species of fish

1. Among the Indian major carps, Catla (*Catla catla*) is the fast growing fish. It consumes a lot of vegetation and decomposing higher plants. It is mainly a surface and column feeder.
2. Rohu (*Labeo rohita*) is a column feeder and feeds on growing plants, decomposing vegetation, large colonial algae, zooplanktons and detritus to a small extent.
3. Calbasu (*Labio calbasu*) is a common feeder on detritus.

4. Mrigal (*Cirrhimus mrigale*) is a bottom feeder, taking detritus to a large content, diatoms, filamentous and other algae and higher plants.
5. Common carp (*Cyprinus cario*) is a bottom feeder and omnivorous.
6. Silver carp (*Hypophthalmichlthys molitrix*) is mainly a surface and phytoplankton-feeder and also feeds on microplants.
7. Grass carp (*Cyernus carpia*) is a specialized feeder on aquatic plants, cut grass and vegetable matter. It is also a fast growing exotic fish.

Poly fish culture: The phytofagous fish (Catla, rohu and mrigal) can be combined with omnivorous (Common Carp), plankton- feeder (silver Carp) and mud eaters (Mrigal and Calbasu) in a composite fish culture system.

Management: The fish are to be nourished with supplementary feeding with rice brands and oilseed cakes. This will enable faster growth and better yield. Each variety of carps could be stocked to 500 fingerlings with the total of 5000-8000 per ha. This stocking density will enable to get a maximum yield of 2000-5000 kg/ha of fish annually.

6. Mushroom cultivation

Mushroom is an edible fungi with great diversity in shape, size and color. Essentially mushroom is a vegetable that is cultivated in protected farms in a high sanitized atmosphere. Just like other vegetables, mushroom contains 90% moisture with high in quality protein. Mushrooms are fairly good sources of vitamin C and B complex. The protein have 60-70% digestibility and contain all essential amino Acids. It is also rich source of minerals like Ca, P, K, Cu. They contain less of fat (0.3%) and CHO and are considered good for diabetic and blood pressure patients. Fibre content in mushrooms is very high which helps in excretion of waste and prevention of constipation.

Types of mushroom

1. Oyster mushroom- *Pleurotus spp*. (var CO1, APK 1, MDU 1,2, Ooty 1)
2. Milky mushroom- *Calsybe indica* (APK 2)
3. Button mushroom- *Agricus gisporus* (Var. Ooty 1)

Oyster mushroom basically occurs in dead woods in nature and resembles the shape of oyster shell. It can be grown on a wide range of agricultural wastes. It can be grown on a wide range of temperature and with low investment. The production of this mushroom is being done on a small scale in different parts of the country and is most popular one.

Milky mushroom is new of its kind and was introduced for commercial cultivation in the year 1998 from TNAU with the variety, APK 2. The mushrooms are umbrella shaped, robust, attractive and milky white in nature. This variety is tropical in nature and can be grown when button and oyster mushroom cannot be grown due to high temperature.

Button mushroom is particularly suitable for winter season, in the northern states. Cultivation of this mushroom requires more investment. This mushroom cannot be cultivated in paddy or wheat straw.

Method of production (Oyster mushroom)

Take fresh paddy straw and cut into small pieces of 3-5 cm length. Soak them in water for 4-6 hours and then boil for half an hour. Drain the water and dry the straw in shade till it is neither too dry nor wet. Take polethene bags of 60X30 cm of size and make two holes of one cm diameter in center of the bag such that they face opposite sides. Tie the bottom of the bag with a thread to make a flat bottom. Fill the bag with paddy straw to 10 cm height. Then incolate with the spawn. Likewise prepare 4-5 layers of straw and spawn alternately. The last layer ends up in the straw of 10 cm heght. Keep this in a spawn running room maintained at a temperature of about 20-25 °C. and with RH 85-90%. After 15-20 days when the spawn running is completed, cut open the polethene bag and take it to cropping room and allow it to grow for 7 days and harvest the mushroom. Yield: 0.5- 1.0 kg/bed.

7. Biogas plant

Biogas is a clean, unpolluted and cheap source of energy, which can be obtained by a simple mechanism and little investment. The gas is generated from the cow dung during anaerobic decomposition. Biogas generation is a complex biochemical process. The celluloitic materials are broken down to methane and carbon dioxide by different groups of microorganisms. It can be used for cooking purpose, burning lamps, running pumps etc.

Types of biogas plant

1. Float dome type: Different models are available in this category, e.g. KVIC vertical and horizontal, Pragati model, Ganesh model.
2. Fixed dome type: The gas plant is dome shaped underground construction. The masonry gas holder is an integral part of the digester called dome. The gas produced in digester collected in dome at vertical pressure by displacement of slurry in inlet or outlet. The entire construction is made of bricks and cement. The models available in this category are janata and deen-bandhu.

The selection of particular type depends on technical, climatologically, geographical and economic factors.

Size: The size of biogas plant is decided by the number of family members and the availability of cattle. One cubic meter capacity plant will need 2-3 animals and 25 kg of dung. The gas produced will meet the requirement of the family of 4-6 members. It would suffice to have a 2 cubic meter plant to cater to the needs of a family of 6-10 members.

Site selection and management: The site should be closed to the kitchen or the place of the use. It will reduce the cost of gas distribution system. It should also be nearer to the cattle shed to reduce the cost of trasportaion of cattle dung. Land should be leveled and slightly above the ground level to avoid inflow or run off of water. Plant should get clear sunshine during most part of the day.

Generation of dung has directed bearing on the quantity of gas generated. The amount of gas production is considerably higher in summerfollowed by rqainy and winter seasons. Gas production would be maximu,m at a temperature between 30 to 35° C. if the amb\ient temperature falls below 10°C, gas production is reduced drastically.

Biogas slurry: Slurry is obtained after the production of biogas. It is an enriched manure. Another positive aspect of this manure is that even after weeks of exposure to the atmosphe are, the slurry does not attract flies and worms.

Factors affecting farming systems

Farmers make decisions about what to grow, what animals to keep, the level and type of inputs and the methods they will use. Their decisions are based upon a range of social, economic and environmental factors. The farmers' attitudes and level of knowledge are also important.

Social and economic factors

These are human factors and include labour, capital, technology, markets and government (political). The importance of each factor varies from farm to farm.

Labour

- In LEDCs, such as India and Java, farmers use abundant cheap labour instead of machines. In Japan and the UK, where labour is expensive, they use machines.
- People working on farms may be unskilled labourers or skilled and able to use machinery, e.g. tractors, harvesters and milking machines.

Capital (finance)

- Capital, the money the farmer has to invest in the farm, can be used to increase the amount of inputs into the farm, e.g. machinery, fences, seeds, fertiliser and renewing buildings.

- If a farmer can afford to invest capital, yields will rise and can create greater profits which can be used for more investment.

Technology

- Machines and irrigation are two types of technology that can increase yields.
- Greenhouses, with computer-controlled technology, provide ideal conditions for high quality crops. The computer controls the temperature, moisture level and amount of feed for the plants.
- Genetic engineering has allowed new plants to be bred that resist drought and disease and give higher yields.

Markets

- Farmers grow crops which are in demand and change to meet new demands, e.g. rubber plantation farmers in Malaysia have switched to oil palm as the demand for rubber has fallen.
- Markets vary throughout the year and farmers change their production to suit them.

Government

- Governments influence the crops farmers grow through regulations, subsidies and quotas.
- Governments offer advice, training and finance to farmers and, in new farming areas, may build the infrastructure of roads and drainage, e.g. Amazonia.
- In some countries, e.g. Kenya and Malaysia, the government is trying to help nomadic farmers to settle in one place.
- Some governments plan and fund land reclamation and improvement schemes.

Environmental factors

- These are physical factors and include climate, relief and soil.

Climate

- Temperature (minimum 6°C for crops to grow) and rainfall (at least 250mm to 500mm) influence the types of crops that can be grown, e.g. hot, wet tropical areas favour rice, while cooler, drier areas favour wheat.
- The length of the growing season also influences the crops grown, e.g. wheat needs 90 days. Some rice-growing areas have two or three crops per year.

Relief

- Lowlands, such as flood plains, are good for crops.
- Steep slopes hinder machinery and have thinner soils; lower, more gentle slopes are less prone to soil erosion.
- Tea and coffee crops prefer the well-drained soil on hill slopes.
- Temperature decreases by 6.5°C for every 1000 metres gained in height.
- South-facing slopes receive more sunlight.

Soil

- Fertility is important; poor soil means lower outputs or larger inputs of fertilisers.
- Floodplains are good for rice because of the alluvial soils.
- Good drainage reduces the dangers of waterlogging.

Competition from the global market

Wheat farming in East Anglia is influenced by local conditions and global markets. Adverse weather conditions mean the harvest can be delayed and the crops may be ruined. Wheat prices fluctuate in response to supply and demand and prices are set according to global markets. Other parts of the world can produce wheat more cheaply than Britain and so farmers must remain competitive. Science, technology and plant breeding have increased output considerably and British wheat production has trebled.

3

Cropping System and Pattern; Efficient Cropping System and Their Evaluation; Tools for Determining Production Efficiencies in Farming Systems

Terms associated with cropping and farming systems are:

1. **Cropping system:** is an important component of a farming system. It represents cropping pattern used on a farm and their interaction with farm resources, other farm enterprises and available technology which determine their makeup.
2. **Cropping pattern**: means the proportion of area under various crops at a point of time in a unit area. It indicates the yearly sequence and spatial arrangement of crops and fallow in an area.
3. **Cropping scheme** is a plan according to which crops are grown on individual plot of a farm during a given period of time with the object of obtaining maximum return from each crop without impairing soil fertility. Thus a cropping scheme is related to the most profitable use of resources, land, labour, capital, and management.
4. **Multiple cropping:** Growing two or more crops on the same piece of land in one calendar year is known as multiple cropping. It is intensification of cropping in space and time dimensions. It includes intercropping, mixed cropping and sequence cropping.
5. **Quadruple cropping:** Growing of four crops in a year in sequence.
6. **Competition effect:** Competition of intercropped spp. For light, nutrients, water, carbon dioxide, and other growth factors.
7. **Complementary effect:** Effect of one component on another which enhances growth and productivity.
8. **Intercropping:** Intercropping is growing two or more crops simultaneously on the same piece of land with a definite row pattern. For example growing maize + green gram in 2:1 ratio

9. **Mixed cropping**: is growing two or more crops simultaneously intermingled without any row pattern. It is common practice in most of dry land areas
10. **Sequence cropping**: sequence cropping can be defined as growing of two or more crops in sequence on the same piece of land in a farming year.
11. **Parallel cropping:** Cultivation of such crops which have different natural habit and zero competition e.g. Black gram /green gram+maize. The peak nutrient demand period for green gram is around 30-35 DAS while it is 50 DAS for maize.
12. **Multi-storied/multi-tiered cropping/multi-level:** Cultivation of two or more than two crops of different heights simultaneously on a certain piece of land in a certain period e.g. sugarcane+mustard+onion.
13. **Sustainable agriculture** is a form of agriculture aimed at meeting the needs of present generation without endangering the resource base of future generation. It is the practice of farming using principles of ecology, the study of relationships between organisms and their environment. It has been defined as "an integrated system of plant and animal production practices having a site-specific application that will last over the long term.
14. **Contour farming** is the practice of tilling sloped land along lines of consistent elevation in order to conserve rainwater and to reduce soil losses from surface erosion. These objectives are achieved by means of furrows, crop rows, and wheel tracks across slopes, all of which act as reservoirs to catch and retain rainwater, thus permitting increased infiltration and more uniform distribution of the water.Contour farming has been practiced for centuries in parts of the world where irrigation farming is important. Although in the United States the technique was first practiced at the turn of the 19th century,
15. **Crop rotation:** The practice of planting a succession of crops in a field over a period of years. Rotations can maintain field fertility since different crops use different soil nutrients, so excessive demands are not made of one nutrient. In certain rotations, plants like legumes (peas and beans) are grown to restore fertility. Crop rotation is a type of cultural control that is also used to control pests and diseases that can become established in the soil over time. The changing of crops in a sequence tends to decrease the population level of pests. Plants within the same taxonomic family tend to have similar pests and pathogens. By regularly changing the planting location, the pest cycles can be broken or limited. For example, root-knot nematode is a serious problem for some plants in warm climates and sandy soils. It is also difficult to control weeds similar to the crop which may contaminate the final produce. For instance, ergot in weed grasses is difficult to separate from harvested grain.

16. **Agroforestry** is a collective name for land use systems and practices in which woody perennials are deliberately integrated with crops and/or animals on the same land management unit. The integration can be either in a spatial mixture or in a temporal sequence. There are normally both ecological and economic interactions between woody and non-woody components in agroforestry. In agroforestry systems, trees or shrubs are intentionally used within agricultural systems, or non-timber forest products are cultured in forest settings. Knowledge, careful selection of species and good management of trees and crops are needed to optimize the production and positive effects within the system and to minimize negative competitive effects.

17. **Alley cropping:** Agroforestry, farm forestry and family forestry can be broadly understood as the commitment of farmers, alone or in partnerships, towards the establishment and management of forests on their land. Where many landholders are involved the result is a diversity of activity that reflects the diversity of aspirations and interests within the community. Alley cropping, sometimes referred to as 'sun systems', is a form of intercropping.

18. **Organic farming** is the form of agriculture that relies on techniques such as crop rotation, green manure, compost and biological pest control to maintain soil productivity and control pests on a farm. Organic farming excludes or strictly limits the use of manufactured fertilizers, pesticides (which include herbicides, insecticides and fungicides), plant growth regulators such as hormones, livestock antibiotics, food additives, and genetically modified organisms·

19. **Organic agriculture** is a production system that sustains the health of soils, and people. It relies on ecological processes, biodiversity and cycles adapted to local conditions, rather than the use of inputs with adverse effects. Organic agriculture combines tradition, innovation and science to benefit the shared environment and promote fair relationships and a good quality of life for all involved.

20. **Permaculture** is an approach to designing human settlements and agricultural systems that is modeled on the relationships found in nature. It is based on the ecology of how things interrelate rather than on the strictly biological concerns that form the foundation of modern agriculture. Permaculture aims to create stable, productive systems that provide for human needs; it's a system of design where each element supports and feeds other elements, ultimately aiming at systems that are virtually self-sustaining and into which humans fit as an integral part.

21. **Deforestation** is the removal of a forest or stand of trees where the land is thereafter converted to a nonforest use. Examples of deforestation include conversion of forestland to farms, ranches, or urban use. The term *deforestation* is often misused to describe any activity where all trees in an area are removed. However in temperate mesic climates, the removal of all trees in an area—in conformance with sustainable forestry practices—is correctly described as regeneration harvest
22. **Agroecology** is the application of ecological principles to the production of food, fuel, fiber, and pharmaceuticals. The term encompasses a broad range of approaches, and is considered "a science, a movement, [and] a practice.
23. **Forest farming** also known as 'shade systems', is the sustainable, integrated cultivation of both timber and non-timber forest products in a forest setting. Forest farming is separate and distinct from the opportunistic exploitation / wild harvest of non-timber forest products. Successful forest farming operations produce: mushrooms, maple and birch syrup, native plants used for landscaping and floral greenery.
24. **Silvipasture:** Silvipastures combine livestock grazing on forage crops or pastures within actively managed tree or shrub crops. Cattle, sheep and goats are the most common livestock incorporated into silvipasture systems and they may be deployed entirely within a private farm/woodlot silvipasture or through collaborative arrangements between forest licensees and livestock producers on public lands.
25. **Food Circle** is a "dynamic, community-based and regionally-integrated food systems concept/model/vision. In effect, it is a systems ecology. In contrast to current linear production- consumption systems, the food circle is a production-consumption-recycle model. A celebration of cycles, this model mirrors all natural systems and is based on the fact that all stable, biological and other systems function as closed cycles or circles, carefully preserving energy, nutrients, resources and the integrity of the whole." [Ibid]
26. **Low Input Agriculture**: Low input farming systems "seek to optimize the management and use of internal production inputs (i.e. on-farm resources)... and to minimize the use of production inputs (i.e. off-farm resources), such as purchased fertilizers and pesticides, wherever and whenever feasible and practicable, to lower production costs, to avoid pollution of surface and groundwater, to reduce pesticide residues in food, to reduce a farmer's overall risk, and to increase both short- and long-term farm profitability." The term is "somewhat misleading and indeed unfortunate. For some it implied that farmers should starve their crops, let the weeds choke them out, and let insects clean up what was left. In fact, the term low-input referred to

purchasing few off-farm inputs (usually fertilizers and pesticides), while increasing on-farm inputs (i.e. manures, cover crops, and especially management). Thus, a more accurate term would be different input or low external input rather than low-input."

27. **Natural Farming**: Natural Farming reflects the experiences and philosophy of Japanese farmer Masanobu Fukuoka. His books *The One-Straw Revolution: An Introduction to Natural Farming* describe what he calls "do- nothing farming" and a lifetime of nature study. "His farming method involves no tillage, no fertilizer, no pesticides, no weeding, no pruning, and remarkably little labour! He accomplishes all this (and high yields) by careful timing of his seeding and careful combinations of plants (polyculture). In short, he has brought the practical art of working with nature to a high level of refinement."

28. **Nature Farming:** Nature farming grew out of the philosophy and methodology of Japanese philosophist, Mokicho Okada in the mid-1940s. "The theory of Nature Farming, as Okada expounded it, rests on a belief in the universal life-giving powers that the elements of fire, water, and earth confer on the soil. The planet's soil, created over a span of eons, has acquired life-sustaining properties, in accordance with the principle of the indivisibility of the spiritual and the physical realms, which in turn provide the life-force that enables plants to grow. To utilize the inherent power of the soil is the underlying principle of Nature Farming." Practices focus on analyzing and building soil through composting, green manuring, mulch, and various other soil management techniques. Similar in many ways to organic farming, nature farming is most commonly practiced in the Pacific Rim countries of Asia and North America.

29. **Kyusei Nature Farming**: Developed by Teruo Higa in Japan during the 1980s, "Kyusei Nature Farming means saving the world through natural or organic farming methods... An added dimension of Kyusei Nature Farming is that it often employs technology involving beneficial microorganisms as inoculants to increase the microbial diversity of agricultural soils, which, in turn, can enhance the growth, health, and yield of crops."

30. **Nutrient Management**: Nutrient management is "managing the amount, source, placement, form, and timing of the application of nutrients and soil amendments to ensure adequate soil fertility for plant production and to minimize the potential for environmental degradation, particularly water quality impairment." Nutrient management has taken on new connotations in recent times. Soil fertility traditionally dealt with supplying and managing nutrients to meet crop production requirements, focusing on optimization of

agronomic production and economic returns to crop production. Contemporary nutrient management deals with these same production concerns, but recognizes that ways of farming must now balance the limits of soil and crop nutrient use with the demands of intensive animal production. Current decision-making processes include crop and animal production factors, economic factors, and the integrity of local surface water and groundwater, as well as the fate of far-away environmental systems.

31. **Precision Farming/Agriculture:** Precision agriculture is a "management strategy that employs detailed, site-specific information to precisely manage production inputs. This concept is sometimes called Precision Agriculture, Prescription Farming, Site-specific Management. The idea is to know the soil and crop characteristics unique to each part of the field, and to optimize the production inputs within small portions of the field. The philosophy behind precision agriculture is that production inputs (seed, fertilizer, chemicals, etc.) should be applied only as needed and where needed for the most economic production." This system requires the utilization of sophisticated technology including personal computers, telecommunications, global positioning systems (GPS), geographic information systems (GIS), variable rate controllers, and infield and remote sensing. Chemical inputs are reduced in precision agriculture, but several factors make it controversial in the sustainable agriculture community, including the requirements of large capital outlay and advanced technical expertise.

32. **Regenerative Agriculture:** Robert Rodale coined this term, and it subsequently was expanded to "regenerative/sustainable agriculture" by the Rodale Institute and Rodale Research Center. Two reasons given for the emphasis on "regenerative" are (1) "enhanced regeneration of renewable resources is essential to the achievement of a sustainable form of agriculture," and (2) "the concept of regeneration would be relevant to many economic sectors and social concerns."

33. **Sustainable Development:** During the past 20 years, considerable interest in sustainability as applied to all areas of human activity has emerged worldwide. Sustainable development must ... "meet the needs of the presentwithout compromising the ability of future generations to meet their own needs." "The vision is of a life-sustaining Earth. We are committed to the achievement of a dignified, peaceful and equitable existence. A sustainable Nation will have a growing economy that provides equitable opportunities for satisfying livelihoods and safe, healthy, high quality of life for current and future generations. Our nation will protect its environment, its natural resource base, and the functions and viability of natural systems on which all life depends."

34. **Whole Farm Planning:** Whole farm planning strategies share a conservation, family-oriented approach to farm management, although specific components may vary from farm to farm, and from community to community. "Whole farm planning provides farmers with the management tools they need to manage biologically complex farming systems in a profitable manner. As a management system, it draws on cutting-edge management theory used by other businesses, industries and even cities. It encourages farmers to set explicit goals for their operation; carefully examine and assess all the resources — cultural, financial, and natural — available for meeting their goals; develop short- and long-term plans to meet their goals; make decisions on a daily basis that support their goals; and monitor their progress toward meeting goals."

35. **Indigenous Farming Systems**

    i) **Shifting Cultivation:** It refers to farming system in north-eastern areas in which land under natural vegetation (usually forests) is cleared by slash and burn method, cropped with common arable crops for a few years, and then left unattended when natural vegetation regenerates. Traditionally the fallow period is 10-20 years but in recent times it is reduced to 2-5 years in many areas. Due to the increasing population pressure, the fallow period is drastically reduced and system has degenerated causing serious soil erosion depleting soil fertility resulting to low productivity. In north-eastern India many annual and perennial crops with diverse growth habits are being grown.

    ii) **Taungya Cultivation:** The *Taungya* system is like an organized and scientifically managed shifting cultivation. The word is reported to have originated in Myanmar (Burma) and tauang means hill, *ya* means cultivation i.e. hill cultivation. It involves cultivation of crops in forests or forest trees in crop-fields and was introduced to Chittagong and Bengal areas in colonial India in 1890. Later it had spread throughout Asia, Africa and Latin America. Essentially, the system consists of growing annual arable crops along with the forestry species during early years of establishment of the forest plantation. The land belongs to forest department or their large scale leases, who allow the subsistence farmers to raise their crops and in turn protect tree saplings. It is not merely temporary use of a piece of land and a poverty level wage, but is a chance to participate equitably in a diversified and sustainable agroforestry economy.

    iii) **Zabo Cultivation:** Zabo is an indigenous farming system practiced in north eastern hill regions particularly in Nagaland. This system refers to combination of forest, agriculture, livestock and fisheries with well-founded soil and water conservation base. The rain water is collected from the catchment of protected hill tops of above 100% slopes in a pond with seepage control. Silt retention tanks are constructed at several points before the runoff water enters in the pond. The cultivation fully depends on the amount of water stored in the pond. The land is primarily utilized for rice. This system is generally practiced in high altitude hill areas, where it is not possible to construct terraces and or irrigation channels across the slope. This is a unique farming system for food production to make livelihood. Zabo means impounding of water. The place of origin of zabo farming system is thought to be the Kikruma village in Phek district of Nagaland.

Cropping systems for irrigated situations

Following cropping systems are being followed in Haryana and other important states of India. The students should be given home work / assignment about various cropping systems being followed in other states and same should be discussed in next class.

Haryana

S.No.	*Kharif*	*Rabi*	Summer
1.	Cotton	Wheat	-
2.	Rice	Wheat	-
3.	Pearl millet	Wheat	-

Other cropping systems followed in the state are:

S. No.	*Kharif*	*Rabi*	Summer
1.	Pearl millet	Barley	Green gram
2.	Rice	Wheat	Green gram
3.	Rice	Wheat	*Susbania*
4.	Cluster bean	Vegetable	Vegetable
5.	Green gram	Mustard	
6.	Green gram	Mustard + Chicory	
7.	Pearl millet	Wheat (Tall)	Vegetable

Other states

Karnataka

S. No.	*Kharif*	*Rabi*	Summer
1	Rice	Rice	Rice
2	Rice	Rice	Ridge Gourd
3	Rice	Spinach	Black gram
4	Rice	Fenugreek	*Green gram*

Punjab

S. No.	*Kharif*	*Rabi*	Summer
1	Rice	Wheat	-
2	Basmati Rice	*Gobhi Sarson*	*Green gram*
3	Maize	Potato	Maize
4	Maize	Wheat	*Green gram*
5	Maize + Turmeric	Wheat + Linseed	
6	Sorghum + Cowpea	Wheat + Mustard	Cowpea

S. No.	*Kharif*	*Rabi*	Summer
1	Pearl millet	Wheat	-
2	Pearl millet	Wheat	Green gram
3	Rice	Wheat	-
4	Rice	Wheat	Green gram
5	Maize	Wheat	-
6	Maize	Potato	-
7	Maize	Potato	Onion
8	Rice	Wheat	Okra
9	Maize	Garlic	Green gram
10	Soybean	Wheat	-
11	Pigeon pea	Wheat	-
12	Green gram	Mustard	-
13	Sesbania (GM)	Potato	Okra
14	Pearl millet	Potato	Cluster bean
15	Sorghum (F)	Oat (F)	Cowpea (F)
16	Sesame	Barley	Green gram

Gujrat

S. No.	*Kharif*	*Rabi*	Summer
1.	Peanut	Wheat	Follow
2.	Peanut	Onion	Sorghum (F)
3.	Peanut	Potato	Sesame
4.	Cotton	Cotton (Contd)	Peanut
5.	Sesame	Potato	Maize
6.	Pigeon pea	Pigeon pea (contd)	Pearl millet
7.	Onion	Chickpea	Maize (F)
8.	Sorghum (F)	Castor	Castor
9.	Sorghum	Cumin	Green gram
10.	Onion	Wheat	Green gram

11.	Castor	Castor (contd.)	Pearl millet
12.	Castor	Castor (contd.)	Green gram
13.	Green gram	Castor	Castor (Contd.)
14.	Peanut	Amaranths	Cowpea
15.	Cotton	Cotton (Contd.)	Pearl millet
16.	Cotton + Green gram	Cotton (Contd.)	Pearl millet
17.	Cotton	Cotton (Contd.)	Groundnut

Rajasthan

S. No.	*Kharif*	*Rabi*	Summer
1.	Pearl millet	Wheat	
2.	Cluster bean	Mustard	
3.	Green gram	Mustard	
4.	Peanut	Wheat	
5.	Cluster bean	Pea	
6..	Pearl millet	Fenugreek (Green leaves)	
7.	Soybean	Wheat	
8.	Maize	Mustard	Green gram
9.	Maize	Chickpea	Cowpea
10.	Maize	Garlic	
11.	Cotton	Cotton (contd.)	Green gram

Andhra Pradesh

S. No.	*Kharif*	*Rabi*	Summer
1.	Maize	Peanut	
2.	Maize	Sunflower	
3.	Maize	Castor	
4.	Maize	Wheat	
5.	Soybean	Maize	
6.	Soybean	Wheat	
7.	Soybean	Castor	
8.	Soybean	Sunflower	
9.	Rice	Rice	
10.	Rice	Maize	
11.	Rice	Sunflower	
12.	Sunflower	Maize	

Cropping systems for dryland situations

For Haryana situations

Mono-cropping in normal & below normal rainfall years:

1. Fallow – rabi crop

Double cropping in above normal rainfall years:

1. Pearl millet – Mustard
2. Green gram – Mustard
3. Cowpea – Mustard

4. Pearl millet + cowpea – mustard
5. Pearl millet + Green gram (8:4) at 30 cm row spacing
6. Pearl millet + Cowpea (8:4) at 30 cm row spacing.

Different locations in the country:

Rice	Lentil	U.P
Rice	Horse gram	Phulbani
Rice	Toria	Biswanath Chariali
Rice	Lentil	
Rice	Mustard	
Rice	Chickpea	
Rice	Potato	
Guava + Sesame		Varanasi
Aonla + Green gram		
Maize	Wheat	Ballowal saunkhuri
Maize	Wheat	Rakh Dhansar
Maize	Black gram	Arjia
Cotton + sesame		Rajkot
Peanut + Castor		
Soybean	Potato	Indore
Soybean	Chickpea	
Sweet Maize	Sweet maize	
Chickpea + mustard in different combinations	1:1	Indore
	2:1	
	3:1	
	2:2	
	4:2	
	6:2	
Chickpea + mustard (mix)		
Pigeon pea + Green gram	1:2, 90x20 cm	Bijapur
	2:1, 120x20 cm	
	2:2, 135x30 cm	
Pigeon pea + Pearl millet	1:3	Solapur
	1:2	
Pigeon pea + cowpea	1:3	
Pigeon pea + Peanut	1:3	
Bt. Cotton + Green gram	1:1, 1:2	S.K. Nagar
Bt. Cotton + Black green	1:1, 1:2	
Bt. Cotton + Moth bean	1:1, 1:2	
Castor + Cowpea		
Maize + Cowpea		

Multiple Cropping Systems

Multiple Cropping or Poly cropping: It is a cropping system where two or three crops are gown annually on the same piece of land using high input without affecting basic fertility of the soil.

Growing two or more crops on the same piece of land in one calendar year known as multiple cropping. It is the intensification of cropping in time and space dimensions i.e. more number of crops within a year and more number of crops on the same piece of land at any given period. It includes inter-cropping, mixed cropping and sequence cropping.

Multiple cropping is a philosophy of maximum crop production per acre of land with minimum of soil deterioration.

1. Rice-potato-green gram.
2. Rice-mustard-maize.
3. Rice-potato-sesame.
4. Jute-rice-potato.

Cropping intensity is more that 200 per cent when the farm as a whole is considered; the Multiple Cropping Index (MCI) is determined by the number of crops and total area planted divided by the total arable area. When the value is three or more, it is said to be most promising farm. This is also called as intensive cropping.

Poly culture: Cultivation of more than two types of crops grown together on a piece of land in a crop season. e.g.

1. Subabul + Papaya + Pigeon pea + Dinanath grass.
2. Mango + Pine apple + Turmeric
3. Banana + Marigold + berseem.

Relay Cropping: Growing the succeeding crop when previous crop attend its maturity stage-or-sowing of the next crop immediately after the harvest of the standing crops. Or it is a system of cropping where one crop hands over land to the crop in quick succession. e.g.

1. Paddy-lathyrus
2. Paddy-Lucerne.
3. Cotton-Berseem.
4. Rice-Cauliflower-Onion-summer gourds.

Overlapping Cropping: In this system, the succeeding crop is sown in the standing crop before harvesting. Thus, in this system, one crop is sown before the harvesting of preceding crops. Here the lucre and berseem are broadcasted in standing paddy crop just before they are ready for harvesting.

Advantages

1. Minimum tillage is needed for relay cropping and primary cost of cultivation is less.

2. Weed infestation is less, as land is engaged with crops year round.
3. Crop residues are added in the soil and thus more organic matter.
4. Residual fertilizer of previous crops benefits succeeding crops.

Efficient cropping system and their evaluation tools

Indices in Cropping System

1. **Land Equivalent Ratio:** It denotes relative land area under sole crop required to produce the same yield as obtained under a mixed or an intercropping system at the same level of management. It is the ratio of land required by pure crop to produce the same yield as intercrop.

   LER = Ya/Sa + Yb/Sb

   Ya, Yb is the yield of a and b crop grown as intercrop,

   Sa, Sb is the yield of a and b crop grown as sole crop,

   LER = Yield of intercrop over yield of pure crop.

2. **Relative Crowding Coefficient (RCC):** It is used in replacement series of intercropping .It indicates whether a crop, when grown in mixed population, has produced more or less yield than expected.

   K_{ab}= Yab/Y_{aa_---} Y_{ab} X Z_{ba}/Z_{ab}

   Where, K_{ab}=RCC of crop a intercropped with crop b,

   Y_{ab}=Yield per unit area of crop a intercropped with crop b,

   Y_{aa}= Yield per unit of sole crop a

   $Z_{ab=}$Proportion of intercropped area initially allocated to crop, a

   $Z_{ba=}$Proportion of intercropped area initially allocated to crop, b

   RCC > 1 means yield advantage

   RCC = 1 no difference

   RCC < 1 yield disadvantage

3. **Aggressivity:** It is the mixture of how much the relative yield increase in component a is greater than that for b.

   Aab = Yab / (Yaa x Zab) - Yba/(Ybb x Zba)

   Aab = Zero mean component crops are equally competitive,

   Aab = negative means dominated,

   Aab = Bigger value either positive or negative means bigger difference in competitive abilities.

4. **Competition Index**: It is measure to find out the yield of various crops when grown together as well as separately. It represents the yield per plant of different crops in mixture and their respective pure stand on unit area basis.

   CI= (Yaa-Yab) X (Ybb-Yba) / Yaa x Ybb

   Yab- mixture yield of a crop grown with b

   Yba- mixture yield of b crop grown with a

   Yaa-yield in pure stand of crop a

   Ybb-yield in pure stand of crop b

5. **Competition coefficient**: Ratio of the RCC of any given spp. In the mixture

   CC = RCC of a given spp. /Total RCC of all crops in mixture

6. **Rotational Intensity**: This is calculated by counting the number of crops grown in a rotation and is multiplied by 100 and then divided by the duration of rotation.

7. **Cropping intensity**:

   Cropping intensity = Total cropped area over net cultivated area x 100

   Or area under kharif + rabi + zaid over area under actual cultivation x 100

Other important indices are

Some of the important indices to evaluate the cropping systems are as below:

Land Use Efficiency or Assessesment of Land Use

The main objective is to use available resources effectively. Multiple cropping which include both inter and sequential cropping has the main objective of intensification of cropping with the available resources in a given environment. Several indices have been proposed to compare the efficiencies of different multiple cropping system in turns of land use and these have been reviewed.

1. Multiple Cropping Index or Multiple Cropping Intensity (MCI)

It is the ratio of total area cropped in a year to the land area available for cultivation and expressed in percentage.

$$MCI=\frac{a1}{A}\times 100$$

Where = 1, 2, 3, n, n= total number of crops, ai = area occupied by crop and A= total land area available for cultivation. Or MCI is the sum of area planted to different crops and harvested in a single year divided by total cultivable area and expressed as percentage. Or MCI means the sum of areas under various crops raised in a single years divided by net area available for that cropping pattern multiplied by 100. It is similar to cropping intensity.

$$\text{MCI}=\frac{\text{Total number of crops + with their respective area}}{\text{Net cultivable area}}\times 100$$

2. Cultivated Land /Utilization Index (CLUI)

Cultivated land utilization Index is calculated by summing the products of land area to each crop, multiplied by the actual duration of that crop divided by the total cultivated land times 365 days.

$$\text{CLUI}=\frac{\text{a1d1}}{\text{A}\times 365}\times 100$$

Where, I = 1, 2, 3, n, n = total number of crops. a1 = area occupied by the ith crop, di = days that the ith crop occupied ai and A = total cultivated land area available for 365 days.

CLUI can be expressed as a fraction or percentage. This gives an idea about how the land area has been put into use. If the index is 1 (100%), it shows that the land has been left fallow and more than 1, tells the specification of intercropping and relay cropping. Limitation of CLUI is its inability to consider the land temporarily available to the farmer for cultivation.

3. Crop Intensity Index (CII)

Crop intensity index assesses farmers actual land use in area and time relationship for each crop or group of crops compared to the total available land area and time, including land that is temporarily available for cultivation. It is calculated by summing the product of area and duration of each crop divided by the product of farmers total available cultivated land area and time periods plus the sum of the temporarily available land area with the time of these land areas actually put into use (Menegay et al. 1978). The basic concept of CLUI and CII are similar. However, the latter offers more flexibility when combined with appropriate sampling procedure for determining and evaluating vegetable production and cropping pattern data.

$$\text{CII}=\frac{\text{ai ti}}{\text{AOT+A1T1}}$$

Where, I = 1, 2, 3……NC, NC = total number of crops grown by a farmer during the time period. T, ai – area occupied by ith crop (months that the crop I occupied an area ai) T = time period under study (usually one year), AO= Total cultivated land area available with the farmer for use during the entire time period, T, M– total number of fields temporarily available to the farmer for cropping during time period. T, j=1, 2, 3........ M, Aj=land area of jth field and Tj= time period Aj is available. When, CII = 1 means that area or land resources have been fully utilized and less than 1 indicates under of resources.

CII indicates the number of times a field is grown with crops in a year. It is calculated by dividing gross cropped area with net area available in the farm, region or country multiplied by 100.

$$\text{CII}=\frac{\text{Gross cropped area}}{\text{Net area}}\times 100$$

When long duration crop grown crop remain longer time in field this is the drawback of CII. So time is not considered, thus , when long duration crops like sugarcane and cotton are in grown, the cropping intensity will be low,though crop remain longer period of time in the field.

4. Specific Crop Intensity Index

It is a derivative of CII and determines the amount of area – time denoted to each crop or group of crops compared to total available to the farmers.

$$\text{SCII}=\frac{\text{ak.tk}}{\text{AOT} + \text{al tl}}$$

Where Nk= total number of crops within a specific designation such as vegetable crops or field crops grown by the farmer during the time period T. AK = area occupied by the k_{th} crop.

Ik= duration of ki^{th} crop.

Using this formula vegetable intensity index, rice intensity index, field crops intensity index etc.

5. Diversity Index (DI)

It was suggested by Strought (1975). It measures the multiplicity of crops or farm products which are planted in a single year by computing reciprocal sum of squares of the share of gross revenue received from each individual farm enterprises in a single year.

$$\text{DI}=\frac{1}{\frac{\text{Yi}}{\text{Yi}}}$$

Where,

N = total number of enterprises (crops or farm products) and yi = gross revenue of ith enterprises produced within a year.

6. Harvest Diversity Index (HDI)

It is computed using the same equation as the DI expects that the value of each farm enterprises is replaced by the value of each harvest.

$$HDI=\frac{1}{\frac{Yi}{Yi}}$$

Where,

Yi = gross value of ith crop planted and harvested within a year.

7. Simultaneous Cropping Index (SCI)

It is computed by multiplying the HDi with 10,000 and dividing the product by MCI. (Strought, 1975).

$$SCI=\frac{HDI}{MCI}\times 10,000$$

8. Relative Cropping Intensity Index (RCII)

It is the modification of CII and determines the amount of area and time allotted to 1 crop or groups of crop related to area- time actually used in the production of all crops. RCII numerator equal SCII denominator and RCII denominator equal CII numerator.

$$RCII=\frac{ak.tk}{ai.ti}$$

These indices can be used for classifying (i.e a farmer with relative vegetable intensity index 50% would be considered a vegetable grower.) measuring shifts of various crops among farm of different sizes and determining whether the consistent types of cropping pattern occur within various farm size strata. These indices also held to know how intensity cultivated land, area has been utilized. But none of these indices takes productivity into account and cannot be used for comparing different cropping systems and evaluating their efficiency in utilization of the resources other than the land.

9. Crop Equivalent Yield (CEY)

Many types of crops/ cultivars are included in a multiple cropping sequences. It is very difficult to compare the economic produce of one crop to another. To cite an example, yield of rice cannot be compared with the yield of grain cereals or pulse crops and so on. In such situations, comparisons can be made based on economic returns (gross or net returns). The yield of protein and carbohydrate equivalent can also be calculated for valid comparison, Effort have also been made to convert the yields of different crops into equivalent yield of any one crop such as wheat equivalent yield and evolved the equation for calculating wheat equivalent yield (WEY).

Crop equivalent yields (CEY): The yields of different intercrops are converted into equivalent yield of any one crop based on price of the produce.

Tools for determining production efficiencies in farming systems

1. Productivity (Productivity per unit area): To estimate the productivity of a component and compares with the crop component expressed in terms of equivalent crop yield. Further the production estimation itself varies among the interlinked animal component in IFS i.e. Rice based farming system

   Productivity in term of grain yield can be recorded and expressed as kg of grain equivalent yield (GEY),

   GEY= [Productivity of component/intercrop (kg) x Cost of component/ intercrop (Rs/kg)]/ Cost of main crop (Rs/kg)

2. Economic analysis: Parameters like cost of cultivation/production, gross and net return and per day return can be worked out and expressed as Rs/ha.

3. Employment generation: Labour required for various activities in crop production given as man days/ha/year (A man working for 8 hours in a day is considered as one man day; A woman working for the same period is treated as 2/3 man day and computed to man days).

4. Productivity of Livestock Components: Milk (per day or lactation), dung, urine etc

   Fisheries: Fish weight recorded at harvest and expressed as kg/unit area.

   Poultry: Egg production per day from birds and expressed as total number per month/year from the unit.

   Pigeon: Growth rate at monthly interval and weight at the time of disposal recorded and expressed as kg/unit.

5. Mushroom: Yield per day and total yield per year from the unit.

6. Water requirement: Water requirement for varying component linkages in the IFS expressed in ha-cm.

7. Residue addition: the quantity of residue available from each component (kg). Potential residue addition in terms of N, P and K.

8. Energy efficiency: Energy input and output was worked out for individual components based on the input and output energies and energy efficiency.

9. Nutritive value: Nutritive value in terms of carbohydrates, proteins and fat (kg)

The overall benefits of crop - livestock integration can be summarized as follows:

- Agronomic, through the retrieval and maintenance of the soil productive capacity.
- Economic, through product diversification and higher yields and quality at less cost.
- Ecological, through the reduction of crop pests (less pesticide use and better soil erosion control); and
- Social, through the reduction of rural urban migration and the creation of new job opportunities in rural areas.
- It helps improve and conserve the productive capacities of soils, with physical, chemical and biological soil recuperation. Animals play an important role in harvesting and relocating nutrients, significantly improving soil fertility and crop yields.
- It is quick, efficient and economically viable because grain crops can be produced in four to six months, and pasture formation after cropping is rapid and inexpensive.
- It helps increase profits by reducing production costs. Poor farmers can use fertilizer from livestock operations, especially when rising petroleum prices make chemical fertilizers unaffordable.
- It results in greater soil water storage capacity, mainly because of biological aeration and the increase in the level of organic matter.
- It provides diversified income sources, guaranteeing a buffer against trade, price and climate fluctuations.

Evaluation of FSR/IFS

1. Productivity (Productivity per unit area): To estimate the productivity of a component and compares with the crop component expressed in terms of equivalent crop yield. Further the production estimation itself varies among the interlinked animal component in IFS i.e. Rice based farming system

   Productivity in term of grain yield can be recorded and expressed as kg of grain equivalent yield (GEY),

   GEY= [Productivity of component/intercrop (kg) x Cost of component/ intercrop (Rs/kg)]/ Cost of main crop (Rs/kg)

2. Economic analysis: Parameters like cost of cultivation/production, gross and net return and per day return can be worked out and expressed as Rs/ha.
3. Employment generation: Labour required for various activities in crop production given as man days/ha/year (A man working for 8 hours in a day

is considered as one man day; A woman working for the same period is treated as 2/3 man day and computed to man days).

4. Productivity of Livestock Components: Milk (per day or lactation), dung, urine etc

   Fisheries: Fish weight recorded at harvest and expressed as kg/unit area.

   Poultry: Egg production per day from birds and expressed as total number per month/year from the unit.

   Pigeon: Growth rate at monthly interval and weight at the time of disposal recorded and expressed as kg/unit.

5. Mushroom: yield per day and total yield per year from the unit.
6. Water requirement: Water requirement for varying component linkages in the IFS expressed in ha-cm.
7. Residue addition: the quantity of residue available from each component (kg). Potential residue addition in terms of N, P and K.
8. Energy efficiency: Energy input and output was worked out for individual components based on the input and output energies and energy efficiency.
9. Nutritive value: Nutritive value in terms of carbohydrates, proteins and fat (kg).

4

Allied Enterprises and Their Importance Criteria for Enterprise Selection

The basic points that are to be considered while choosing appropriate enterprise in IFS are:

1. Soil and climatic features of an area/locality.
2. Social status of the family and social customs prevailing in the locality.
3. Economic condition of the farmer (Return/income from the existing farming system).
4. Resource availability at farm and present level of utilization of resources.
5. Economics of proposed IFS and credit facilities.
6. Farmer's managerial skill.
7. Household demand.
8. Institutional infrastructure and technological knowhow.
9. Market facilities.

Priority should be given to improve the present status of different components of the existing farming system, should have better compatibility with prevalent farming system, nil to very less dependence of input from outside, high risk bearing and capable to generate more per day income and employment. In addition to this technological knowhow related to the enterprise(s) should locally be available and particularly no wastage of products and by products due to integration of enterprises.

1. Integration of enterprises

In agriculture, crop husbandry is the main activity. The income obtained from cropping is hardly sufficient to sustain the farm family throughout the year. Assured regular cash flow is possible when crop is combined with other enterprises. Judicious combination of enterprises, keeping in view of the environmental condition of locality will pay greater dividends. At the same time, it will also promote effective recycling of residues/wastes.

The principle of combing enterprises

A farm manager is often confronted with the problems as to what enterprises to select and the level at which each enterprises should be taken up. How far he can go far or should go in combine enterprises with another depends partly on the inter-relationships, between different enterprises and the prizes of products and inputs.

Types of Enterprises

1. Independent enterprises,
2. Joint enterprises,
3. Supplementary enterprises,
4. Complementary enterprises and
5. Competitive Enterprises

Independent Enterprises

Independent Enterprises are those which have no direct bearing on each other, an increase in the level of one neither help not-hinders the level of other. In such cases each product should be treated separately e.g. production of wheat and maize independently.

Joint Enterprises

Joint products are those which are produced together e.g. cotton and cottonseed, wheat and straw etc. the quantity of one product decides the quantity of the other products. In case of joint products there is no economic decision are made with respect to the combination of products and two products can be treated as one.

2. Competitive Enterprises

Competitive enterprises are those which compete for use of the farmers' limited resources; use of resources to produce more of the one necessitates a sacrifice in the quantity of other product. When enterprises are competitive three things determine the exact combination of the product, which would be most profitable.

- The rate at which one enterprises substitute for the other.
- Prices of the products and
- The cost of producing the product.

The rate at which one product substitutes for another is known as the marginal rate of substitution.

The products can

1. Constant rates of substitution.
2. Decreasing rate of substitution
3. Increasing rates of substitution
   e.g. paddy- sorghum, paddy- groundnut

3. Supplementary Enterprises

Two products are said to be supplementary when an increase in the level of one does not adversely affect the production of the other but adds to the income of the farm i.e enterprise which do not complete with each other but adds to the total income. For example, on many small farms dairy enterprise or a poultry enterprise may be supplementary to the main crop enterprises because they utilize surplus family labour and shelter available and perhaps even some feeds, which would otherwise go to waste. Some time enterprises are supplementary for one resource but competitive for another. In such cases the relationship should be treated as one of competitive. Even though they are supplementary to one another in respect of other sources e.g. mixed crops.

4. Complementary Enterprises

Complementary enterprises are those, which add to the production of each other e.g. Berseem and maize crops. Two products are complementary when the transfer of available input for the production of the one product to the production of the other results in increases in the production of both products. Then two crops are complementary enterprises, the use of resources for the two crops result in the increased production of both the crops.

Two enterprises do not remain complementary over all possible combinations. They become competitive at some stages. When both complementary and competitive relationship occurs, the complementary relationship occurs first and then is followed by competitive relationship.

Enterprise integration

Livestock is the best complementary enterprise with cropping, especially during the adverse years. Installation of bio-gas plant in crop-livestock system will make use of the wastes, at the same time provides the valuable manure and gas for cooking and lightning. In a wetland farm there are greater avenues for fishery, duck farming and buffalo rearing. Utilizing the rice straw, mushroom production can be started. Under irrigated conditions (garden lands), inclusion of sericulture, poultry and piggery along with arable crop production is an accepted practice. The poultry component in this system can make use of the grains produced in the farm as feed. Pigs are the unique components that can be reared with the wastes which are unfit for human consumption. In rainfed farming, sheep and goat rearing form an integral part of the landscape. Sericulture can be introduced in rainfed farming, provided the climatic conditions permit it. Agro-forestry (Silviculture and silvi-horticulture) are the other activities which can be included under dryland conditions. In the integrated system, selection of enterprise should be on the cardinal principle that there should be minimal competition and maximum complementary effect among the enterprises.

Considerations for enterprise selection

Every farmer faces the question of what to produce. The selection of enterprises is critical in determining whether or not the goals of your family will be met through farming.

The purpose here is to provide a step by step guideline to the enterprise selection process. The first step is to determine the goals or objectives of your farm family. The next step is to find out whether or not they are feasible. This is accomplished by taking an inventory of the resources available and the resource requirements of each enterprise you are considering. Next, realistically compare the resources required to the resources available and cut back the list of alternatives. Finally, select the enterprises you wish to pursue and design a plan for implementation.

This approach is appropriate for farmers who are already engaged in farming as well as the beginning farmer. It is a helpful exercise to repeat periodically to keep the farm on a forward looking course and to keep all alternatives open. Any business operates in a changing and dynamic society. Opportunities as well as goals change over time making periodic reassessment and adjustments in the business necessary.

You are encouraged to keep written records or notes on each step of the process. It forces you to give clear and concrete answers to each of the questions and will serve as a useful record for making decisions in the future.

Determine your goal

Each individual has their own view of success. Most people will answer yes when asked if they have personal goals. Yet very few people will answer yes if asked if they have ever put these goals to paper.

Most people have the broad objectives of achieving financial security, health and safety, and personal growth. But we also have goals that are more specific and involve some sort of action. These goals should be measurable in some way and have a time frame associated with them. When writing down your goals, also write down the time frame and ways you can measure their achievement. This will help in evaluating the success of your business and in developing an implementation plan.

In some cases the goals of family members may conflict. In other cases the goals may be the same but the means for achieving them may not be compatible. It is important for each person involved in the farm to record their goals individually before the goals of the business as a whole are formulated. You may be pleasantly surprised to find out that the goals of individuals seemingly in conflict may be very similar.

The following is a list of questions that can be used to help develop your list of goals:

- Is your primary reason for farming to maximize income, to have a rural lifestyle, to provide income for family members, or other reasons?
- What other activities are you involved in, and what are the priorities of these activities relative to the farm business?
- Do you want to devote full-time effort to the farm or would you prefer farming to be a part-time activity?
- How much are you willing to be restricted by time and capital demands of your farm business?
- Do you want to eventually transfer the ownership of the farm to a partner or family member?
- Is income from the farm and/or sale of the farm an important part of your retirement plan?
- What is the desired period between initial investment and cash returns?
- Do you want to learn new skills through self-study or formal training?

Inventory your resources

The availability of resources will ultimately limit your choice of enterprises simply because the resource requirements among enterprises vary. A list of resources typically includes land, labor and capital. But there are other factors to consider such as climate, access to information, management skills, and markets.

Access to markets is the most commonly overlooked factor in the enterprise selection process. But in fact it can be your most limiting constraint. Simply because you can grow something does not mean you can sell it. And just because you can sell a product does not mean that it will be profitable. A third possibility is that you will be able to sell a product at a money making price but that you will only be able to sell a limited amount of the product; that is, less than the total amount that you are able to produce.

Consider your market potential carefully. If it is a product that has never been tried before in your area plan to take several years to get established. Be realistic about your cash flow situation and plan accordingly. For each of the areas listed below create a list of the resources available. This will be compared later to the resources required by each enterprise you are considering. A written list will enable you to easily check off the requirements on the enterprise resource requirement list later on.

Physical factors

Land

- How much land do you have available?
- What is the physical profile and topography of the land?
- What is the soil texture, drainage capability and nutrient levels?
- Which types of weeds are growing on the soil?
- Which other crops have been grown on the land?

Climate

- What is the average rainfall in your area and when are the rainy periods?
- When are the first and last frost dates and how much have the actual dates varied historically?
- What are the high and low temperatures for your area and when do they occur?
- What is the average daily temperature?
- What is the day/night temperature variation?
- What is the direction and strength of winds?

Irrigation Water

- Where does your water come from and what is its cost?
- What is the water quality?
- Do you have water rights? Are you within an irrigation district?
- When is irrigation water available to you and in what amount?
- What type of irrigation system do you have?
- What are the differences in cost and efficiencies for alternative systems?

Farm Structures

- What type of buildings do you have on the property and what is their condition?

- Do you have structurally sound fences?
- If you feel you need additional buildings or fences, have you checked into the cost of their construction?

Machinery and Equipment

- What type of farm power machinery do you have?
- What farm implements do you have?
- What is your transportation equipment: truck, pick-up, or trailer? Consider capacity and efficiency.
- Have you considered leasing/renting some equipment?
- What are the possibilities of contracting with custom operators in your area?

Financial factors

- How much capital are you willing/ able to invest?
- Are you able or willing to borrow capital?
- What is your cash flow situation?
- Is a high rate of return on your investment important to you?
- Are you willing to consider risky enterprises?

Management

Personal skills

- Management skills: record keeping, personnel management, budgeting, familiarity with tax and other relevant laws - do you consider these to be adequate?
- What are your mechanical skills?
- Which are your knowledge strong points: plant physiology, animal health, pest management, greenhouse production, etc.?
- Would you prefer handling a diversified farm or would you prefer one or two major enterprises?

Information Access

- Are you familiar with the agricultural information delivery systems?
- Are you able to access the resources of these systems?
- Is sufficient information available for the enterprises in which you are interested?
- Are you willing to learn new skills if they are required?

Labor Factors

- What are your labor needs on a monthly basis?
- Are you planning to use mostly family or mostly hired labor?
- Have you checked out the regulations of the California Labor Law?
- Have you considered the opportunity cost of using your own labor?

Marketing Factors

- Do you have a preferred marketing method? Broker, retailer, direct (roadside stand, farmers market, U-pick), cooperative, contract with processor?
- What is your proximity to various potential markets?
- Have you contacted potential markets for their advice on crop selection?
- How much time are you willing to spend marketing your products?
- Do you have cooling facilities for perishable products?
- Are you familiar with marketing regulations for the enterprises you are considering?

Develop a list of possible enterprises

After identifying your goals and resources, develop a list of possible enterprises. The following set of questions and the list at the end of this publication should help.

- Which enterprises are predominant in your area?
- Are there enterprises which interest you that have been successful in other areas in similar soil and climate conditions (i.e., enterprises that have potential in your area but have not yet been established)?
- What crops or livestock have been raised on your land in the past?
- Which are the enterprise types with which you feel more personally compatible: livestock, field crops, orchard crops, small fruits, vegetables, ornamentals, growing transplants, raising seed?

5

Sustainable Agriculture-Problems and Its Impact on Agriculture; Indicators of Sustainability

Agriculture has been the base of subsistence for human settlements on the planet earth. According to the Food and Agricultural Organization (FAO), people in the developing world where the population increase is very rapid, may face hunger if the global food production does not increase by 50–60%. Contribution from developing countries and developed countries to world production in 1975 was about 38% and 62%, respectively. If yield increase is keeping pace with increasing population, mass hunger can escape. In the pre-independence period, Indian agriculture was usually described as a gamble with monsoons and their failures resulted in widespread famine and misery. In the last few decades, Indian agriculture has made a remarkable progress even with increased population and decreased per capita availability of agricultural land. World population is projected to be over 8 billion by 2025 and nearly 10.5 billion by the end of next century. To maintain the status quo, food production needs to be doubled. Estimates by the FAO and WHO (1992) and the Hunger Project (1991) suggest that around 1 billion people in the world have diets that are 'too poor to abstain the energy required for healthy growth of children and minimal activity of adults'. The causes are complex and it is not entirely the fault of overall availability of food. Modern agriculture begins on the research farm, where researchers have access to all types of inputs for crops at all the appropriate times. But, even the best performing farms cannot match the yields of researchers. For high productivity, farmers need to have access to the whole package – improved seeds, water, labour, capital or credit, fertilizers, pesticides and other developed technology. Many poorer farming households simply cannot adopt the whole package. Very often delivery systems are unable to supply them on time.

In December 1983, the UN General Assembly established the World Commission on Environment and Development. In 1987, on 27th of April, at the queen Elizabeth Hall in London, the Prime Minister of Norway, Mrs. Brundtland, who was also the Chairman of the World Commission of Environment and development, released the publication of "**Our Common Future**" by the World Commission on Environment and Development (WCED) and said:

"Securing our common future will require new energy and openness, fresh insights, and an ability to look beyond the narrow bounds of national frontiers and separate scientific disciplines. The young are better at such vision than we, who are too often constrained by the traditions of former, more fragmented World. We must tap their energy, their openness, their ability to see the interdependence of issues..." She suggested that we must adopt a new paradigm based on a completely new value system. "Our generation has too often been willing to use the resources of the future to meet our own short- term goals. It is a debt we can never repay. If we fail to change our ways, these young men and women will suffer more than we, and they and their children will be denied their fundamental right to a healthy productive, life-enhancing environment." Her speech made it clear that we are consuming resources, which must be transferred to the next generation. We must recognize that, because resources are limited, we need a sustainable way of life.

Almost at the same time the realization of prime importance of staple food production for achieving food security for future generations has brought the concept of "Sustainable Agriculture" to the forefront and began to take shape in the following three points.

1. The interrelatedness of all the farming systems including the farmer and the family.
2. The importance of many biological balances in the system.
3. The need to maximize desired biological relationships in the system and minimizes the use of materials and practices that disrupt these relations. Sustainability of agricultural systems has become a global concern today and many definitions so Sustainable Agriculture has become available.

Definition of Sustainable Agriculture

Sustainable agriculture has been defined in many ways by different workers. Some of the definitions are given below:

It is the successful management of resources for agriculture to satisfy changing human needs while maintaining or enhancing the quality of environment and conserving natural resources.

A sustainable Agriculture is a system of agriculture that is committed to maintain and preserve the agriculture base of soil, water and atmosphere ensuring future generations the capacity to feed themselves with an adequate supply of safe and wholesome food'.

'A Sustainable Agriculture system is one that can indefinitely meet demands for food and fibre at socially acceptable, economic and environment cost'.

Some other designated it as regenerative agriculture or alternative farming. Sustainable agriculture is a food and fiber production and distribution system that:

- Supports profitable production
- Protects environnemental quality
- Uses natural resources efficiently
- Provides consumers with affordable, high-quality products
- Decreases dependency on nonrenewable resources
- Enhances the quality of life for farmers and rural communities, and
- Will last for generations to come.

Sustainable Agriculture refers to a range of strategies for addressing many problems that effect agriculture. Such problems include loss of soil productivity from excessive soil erosion and associated plant nutrient losses, surface and ground water pollution from pesticides, fertilizers and sediments, impending shortages of non- renewable resources, and low farm income from depressed commodity prices and high production costs. Furthermore, "Sustainable" implies a time dimension and the capacity of a farming system to endure indefinitely.

A broad and commonly accepted definition of sustainable Agriculture is as follows:

Sustainable Agriculture refers to an agricultural production and distribution system that:

- Achieves the integration of natural biological cycles and controls
- Protects and renews soil fertility and the natural resource base
- Reduces the use of non-renewable resources and purchased (external or off-farm) production inputs
- Optimizes the management and use of on- farm inputs
- Provides on adequate and dependable farm income
- Promotes opportunity in family farming and farm communities, and
- Minimizes adverse impacts on health, safety, wildlife, water quality and the environment

Sustainable v/s Modern agriculture

Particulars	Sustainable Agriculture	Modern Agriculture
Plant nutrients	FYM, Compost, GM, bio-fertilizer, Crop Rotations	Chemical fertilizers
Pest control	Cultural methods, Crop rotations and biological methods	Toxic chemicals are used
Inputs	High diversity, renewable and biodegradable inputs	High productivity and low diversity
Ecology	Stable ecology	Fragile ecology
Use of resources	The rate of extraction from forests, fisheries, underground water resources and other renewable resources don't exceed the rate of regeneration.	The rate of extraction exceeds the rate of regenerate falling of trees, deforestation, over grazing, and pollution of water bodies take place.
Quality of food material	Food materials are safe	Food materials contain toxic residue

Current concept of sustainable agriculture

The ultimate ends of sustainable agriculture is to develop farming systems that are productive and profitable, conserve the natural resource base, protect the environment, and enhance health and safety, and to do so over the long-term. The means of achieving this is low input methods and skilled management, which seek to optimize the management and use of internal production inputs (i.e., on-farm resources) in ways that provide acceptable levels of sustainable crop yields and livestock production and result in economically profitable returns. This approach emphasizes such cultural and management practices as crop rotations, use of animal, green manures and organic wastes and conservation tillage to control soil erosion and nutrient losses and to maintain or enhance soil productivity.

Low-input farming systems seek to minimize the use of external production inputs (i.e., off-farm resources), such as purchased fertilizers and pesticides, wherever and whenever feasible and practicable, to lower production costs: to avoid pollution of surface and groundwater, to reduce pesticide residues in food: to reduce a farmer's overall risk, and to increase both short-term and long-term farm profitability. Another reason for the focus on low- input farming systems is that most high input systems, sooner or later, would probably fail because they are not either economically or environmentally sustainable over the long-term. There are three R's of sustainability-Renewability, Reversibility and Resilience. The resources are renewable and non-renewable. Water from snowfall running into rivers is renewable which is replenished quickly without creating pollution. But coal energy is non-renewable resource which once used takes million years to replenish. Reversibility is the potential of a natural resource to attain pre-use status. Water in an aquifer will replenish if pumping is stopped or postures will attain its greenery if grazing is stopped. Resilience or resistance is the ability of an ecosystem to recover completely from perturbation. Sustainability, therefore, asserts the use of renewable sources of energy as its product will be eco-friendly which can be used many times without worrying about pollution.

Goals of Sustainable Agriculture

A sustainable Agriculture, therefore, is any system of food or fibre production that systematically pursues the following goals:

- Thorough incorporation of natural processes i.e. nutrient cycling, nitrogen fixation and pest-predator relationships into agricultural production processes
- Reduction in the use of those off-farms, external and non-renewable inputs which damage the environment or harm the health of farmers and consumers, and more use of leftover inputs to minimize variable costs
- Full participation of farmers and rural people in all processes of problem analysis and technology development, adoption and extension
- Equitable access to predictive resources and opportunities, and progress towards more socially just forms of agriculture
- More productive use of the biological and genetic potential of plants and animal species
- More productive use of ITK and practices, including innovation in approaches not yet fully understood by researchers or widely adopted by farmers
- Developing self-reliance among farmers and rural people
- Ensuring long-term sustainability through good matching between cropping patterns and productive potential and environmental constraints of climate and landscape
- Profitable and efficient production with integrated farm management and conservation of soil, water, energy and biological resources
- Enhancing compatibility with social and political conditions
- Minimizing adverse environmental impacts on adjacent and downstream environments.

Elements of sustainability

There are a number of ways to improve the sustainability of a farming system and these vary from region to region, However, there are some common sets of practices among farmers for sustainable approach through higher use of on-farm or locally available resources which contribute to long- term profitability, environmental safety and rural quality of life.

a) **Soil conservation-** Many soil conservation practices like contour cultivation, contour bunding, graded bunding, vegetative barriers, strip cropping, cover cropping, reduced tillage etc reduce soil loss due to wind and/or water erosion

b) **Crop diversity-** Growing of different crops on a farm reduces risks from extremes weather, market conditions or crop pests. Increased diversity of crops and other plants, such as trees and shrubs help in soil conservation, wildlife habitat and built up of beneficial insects

c) **Nutrient management-** Judicious management of plant nutrients improves the soil and protects environment. Higher use of on-farm nutrients such as manures and leguminous cover crops cut cost of purchased fertiliser.

d) **Integrated pest management (IPM)-** IPM is a sustainable approach for managing pests by integrating various methods of pest control i.e. biological, cultural, physical and chemical which minimize economic, health and environmental risks.

e) **Cover crops-** Growing cover crops such as cowpea, sun hemp, horse gram in the off season after harvesting a grain or vegetable crop is helpful in weed suppression, erosion control and improved physico-chemical soil properties.

f) **Rotational grazing-** New management- intensive grazing systems take out animals to pasture for providing high-quality forage and reduce feed cost.

g) **Water quality & water conservation-** Various practices have been developed to conserve and protect quality of drinking and surface water namely deep ploughing, mulching, micro irrigation techniques etc..,

h) **Agro forestry-** Trees and other woody plants are often underutilized and cover a range of practices viz., agri-silviculture, silvi-pastoral, agri-silvi-horticulture, horti/silvipastoral, alley cropping, tree farming , lay farm which conserve soil and water.

i) **Marketing-** Farmers across the country are finding improved marketing - way to increase profit. Direct marketing of agricultural product from farmers to consumers is becoming much more common through road side market.

Status of Sustainable Agriculture in India

Overall prosperity of the nation depends on sustainable development. It is a process of social and economic upliftment that satisfies needs and values of interest groups. Suitable development of India demands access to 'clean' technologies and has strategic role in increasing abilities of the country both to the environment and to provide thrust towards conservation and sustainable agriculture. Current research programmes towards sustainable agriculture are as follows:

1. Resistant crop varieties to edaphic, climatic and biotic stresses
2. Multiple cropping systems for irrigated areas and tree based farming system for rainfed areas.
3. Integrated nutrient management
   a. Integration of organic and inorganic sources of nutrients
   b. Growing of green manure crops (*Sesbania, Crotalaria* etc)
   c. Inclusion of legume crops in crop rotation
   d. Use of bio-fertilizers
4. Integrated pest management
   a. Microbial control
   b. Use of bio-agents
   c. Use of predators
   d. Bio-herbicides
5. Soil and water conservation
   a. Watershed management
   b. Use of organics as mulch and manure
   c. Use of bio-fencing like Opuntia spp (Prickly pear)
6. Agro forestry systems in dry lands/ sloppy areas and erosion prone areas
7. Farm implements to save energy in agriculture
8. Use of non-conventional energy in Agriculture
9. Input use efficiency
   a. Water technology
   b. Fertiliser technology
10. Plant genetic resource collection and conservation.

Limitations of sustainable agriculture and its impact on agriculture

1. Yield stagnation

After enjoying the fruits of Green Revolution during last 3 decades the high productive areas are encountering sustainability problems and there are reports of declining factor productivity. In some of the long-term experiments grain yield especially of rice is declining. Stagnation in the crop productivity combined with possible deterioration in soil health has raised doubt on the long-term consequences of the existing practices.

2. Soil degradation

Soil degradation means that the soil loses its natural fertility for the production of food and regenerative raw materials. The most widespread phenomenon is the losses organic matter and of the nutrients needed for biomass production as a result of improper soil management together with wind and water erosion. There is also increasing concern about the deterioration of soil quality in large area due to salinization, acidification and increasing concentration of heavy metals and other persistent substances in the soil. Further, crop production under intensive cultivation over the years has resulted in large scale removal of nutrients from the soil. Inadequate replenishment over years has resulted in negative nutrient balance.

3. Environmental deterioration

Among many concerns, pollution of surface and underground water due to fertilizer has received greater attention. A sizeable amount of nitrogen could escape to the environment as ammonia by volatilization from soil surface, nitrous oxide or elemental nitrogen by leaching in underground water. The ammonia going to the atmosphere contributes to acid rains, while N_2O is involved in depletion of ozone layer. Thus, excessive and imbalanced use of chemical fertilizers and pesticides has not only deteriorated soil fertility but has also poisoned the soil, water, microflora, atmosphere, human beings and other animals, causing several deformities, inabilities and serious diseases.

4. Low soil organic carbon

India's share in overall soil organic carbon stock of the world is only about 3% although it covers 11.9% of total geographical area of the world. This is mainly for heavy deforestation and faulty land-use pattern. Additionally, the unfavourable climatic conditions in the Indian peninsula have further enhanced the rate of decomposition of soil organic carbon and consequent depletion.

5. Low fertility status

Indian soils are generally poor in fertility as they are low in organic matter and have consistently been depleted of their nutrient resource due to continuous cultivation for many centuries. Soils of our country are universally deficient in all nutrients except in some parts of north-eastern region. Nearly 50% soils are deficient in P and 20% in K. Sulphur has become critical on low organic matter coarse-textured soils under S-exhaustive oilseed-based cropping systems. Among the micronutrients, zinc deficiencies are widespread across the country, particularly in coarse-textured lowland soils with low organic matter. The magnitude of boron deficiencies is next to Zn. Appearance of Mn deficiencies has also been observed on coarse-textured highly percolation soil of north-western Indo-Gangetic alluvial plains in rice-wheat/ berseem cropping system.

6. Nutrient mining

Since last 25 years the Indian soils are experiencing, on an average, a net negative balance @ 8-10 mt of nutrients per annum. During 1998-99 alone about 30 mt plant nutrients were removed while only 16.8 mt of fertilizer nutrients added. About 70% of the total gross cropped area in the country experienced a nutrient depletion of more than 50 kg/ha annually. Almost 50% of the nutrient removal is accounted for by potash, whereas its use hardly exceeds 6% it shows wide disparity between nutrient removed and application.

7. Regional disparity in fertilizer consumption

There are a lot of disparities in the fertilizer consumption in different region. Of the total 525 districts of India, 19 districts consume more than 200kg NPK/ha, 75 between 100 and 150 and 132 between 50 and 100. Further, 84 districts account for 50% of the total consumption. Consumption in western and eastern zones are far below the national average of 89.9 kg NPK/ha.

8. Imbalanced fertilizer use

During 1998-99, consumption of N, P_2O_5 and K_2O was 11.32, 4.10 and 1.33 million tonnes, respectively. Imbalanced use of nutrients is reflected from the fact that N: P_2O_5: K_2O consumption ratio widened from 5.9: 2.4:1.0 during 1991-92, the year during which the phosphatic and potassic fertilizers were decontrolled to 8.5: 3.1: 1 during 1998-99. Data on nutrient uptake show that N, P and K are removed in ratio of 5:1: 5, respectively. It shows wide disparity between the ratio of nutrient removed and ratio of nutrient application. Thus, the present soil fertility maintenance approach itself indicates inherent un-sustainability in the management of the definite nutrient reserves. Continuous application of N fertilizers alone causes sharp reduction in soil organic matter, a key indicator of soil quality. Decline in soil organic matter is arrested by balanced application of the NPK in conjunction with annual appli-cation of 10-15 tonnes FYM/ha. Negative response to N application on highly P-deficient soils is a typical example, applied-fertilizer induced unsustainability in crop production.

9. Appearance of diseases, pest and weed hazards

High-yielding crops are generally more susceptible to pests, diseases and other environmental factors unless they are specifically bred for the purpose. With the change in varieties and bio-types adopted on a large scale, pest and disease patterns change. Weed complex also keeps changing constantly.' For example, if we look back about 3 decades, Pohli (*Carthamus oxycantha*) and Piazi (*Asphodelus tenuifolius*) were the two most common obnoxious weeds in the wheat crop in the northern and north-west India. Due to the increased intensity of cropping (rice followed by wheat) in these areas, these weeds are now capacious by their absence. Instead *Phalaris minor* and wild oats have become the most common weeds. Similarly, diseases and pests have changed in their mix, occurrence patterns and virulence. The resistance and virulence patterns of disease and pest problems have changed to render traditional control measures ineffective in several cases.

Strategies for sustainable agriculture

1. Balanced fertilization

Balanced fertilization does not mean a certain definite proportion of nitrogen; phosphorus and potash (or other nutrients) to be added in the form of fertilizers, but it has to be taken into account the availability of nutrients already present in the soil, crop requirement and other factors. It should not mean that every time a crop is grown, all the nutrients should be applied in a particular proportion; rather fertilizer application should be tailored to the crop needs keeping in view the capacity of these soils to fulfil these needs. There is, therefore, a need to work out balanced fertilization formulae for each crop/cropping system in a specific region.

2. Integrated plant nutrient management

Mitigated plant nutrient supply system (IPNS) primarily relates to combined application of organic and inorganic sources of plant nutrients. Organic sources, when applied with mineral fertilizer, improve the efficiency of latter due to their favourable effects on physical and biological properties of soil. The IPNS is the way to get the best out of the both and to assure sustained agriculture.

3. Organic manures

Organic manures not only regularly supply the macro, micro and sec-ondary nutrients but also improve soil physical properties and soil biological health. A 9-year study at Pantnagar showed that response of maize to N declined to zero, while that to P increased from 22 to 42%, to K from 0 to 35 and to FYM from 7 to 35; continuous application of FYM in conjunction with NPK resulted in yield increase of 1 tonne/ha. A field study at Ludhiana revealed that 12 tonnes FYM and 80 kg N/ha produced same yield of rice as 120 kg N/ha and in addition gave residual effects on suc-ceeding wheat equivalent to 30 kg/ha each of N and P_2O_5. Furthermore, there was a build up in soil fertility with FYM application.

4. Green manuring and legume residues

Importance of green manuring in crop production and soil fertility build up is undisputed. When used before a crop, it can bring a saving of about 60 kg N/ha which is half the dose recommended generally for rice or wheat. However, the practice has not become popular because the crop does not give economic returns to the farmers. An alternative to this is growing a summer legume green gram crop, picking matured pods and incorporating the residue; this may be referred to as partial green manuring. A recent study at IARI revealed that benefits of incorporation of green gram residue were similar to that of *Sesbania* green manuring. The agronomic efficiency (kg grain/kg urea- N) was 9.6 for pre-rice fallow, 11.7 for *Sesbania* green manuring and 12.7 for green gram residue incorporation, indicating that incorporation of green manure or residue had a synergistic effect on the response of rice to nitrogen, probably due to favourable effect of organic matter on physical and biological properties of soil. So far, only N effects of green manuring have been measured and reported but recent studies at IARI revealed that incorporation of green manure of green gram residue resulted in an increase in available P content of soil in addition to a significant increase in organic carbon and total N.

5. Cereal residue

Recently, a large area coming under rice-wheat cropping system has made million tonnes of cereal residue surplus on the farm, particularly the rice residue, which has no taker and currently most of it is burnt. Its incorporation in soil is being suggested but has the problem of immobilization of soil and applied nutrients. Incorporation of wheat residue before rice resulted in production of less yield than removal of wheat residue at 0 and 60 kg N/ha but higher yield with 180 kg N/ha. Also application of one-third dose of nitrogen at the time of residue incorporation about one month before the rice transplanting and remaining two-third dose of nitrogen in 2 equal splits at 10 days after transplanting and panicle initiation gave significantly higher yield than application in 2 splits.

6. Legumes in cropping system

Growing legume crops in a cropping system is a well known practice for restoring soil fertility. Legumes leave a substantial amount of residual N (other effects have not been really researched) that may vary from 30 to 60 kg N/ha in the case of grain legumes and 90 to 120 kg N/ha in the case of forage legumes such as ber-seem, lucerne and alfalfa. For example, in a study at IARI, growing of green gram in rainy season contributed to the fertilizer equivalent of 35-56 kg N/h contribution increased to fertilizer equivalent of 74-94 kg N/ha as urea.

7. Bio-fertilizers

Bio-fertilizers are microbial cultures that fix atmospheric nitrogen and include *Rhizobium*, *Azotobacter, Azospirillum*, blue-green algae (BGA) and azolla. In addition to these there are bacterial/fungal cultures that help in phosphate solubilization of both native and applied sparingly soluble phosphates. There has also been growing interest in VAM fungi. Of these biofertilizers *Rhizobium* cultures have been most successful and are recommended for use with legumes. Of late, there has been considerable interest in A*zotobacte*r and Azospirillum for cereals other than rice and in BGA and azolla for rice. The amount of contribution generally attributed to these cultures vary from 10 to 30 kg N/ha. In rice BGA has been widely tested in field experiments. A study conducted at IRRI indicated that when effective, BGA inoculation increased the grain yield of rice by 14% over control. In a study at IARI application of BGA along with 60 kg N/ha gave a yield increase equivalent to 30 kg N/ha applied as urea. Azolla has also received atten-tion from researchers in rice. Data averaged from 55 trials under All-India Co-ordinated Rice Improvement Project show that green manuring with 6 tonnes azolla/ha plus 25 kg urea N produced 94% of the yield obtained by applying 50 kg N/ha.

A group of heterotrophic micro-organisms are known to have the ability to solubilise native or applied sparingly soluble phosphates such as ground rock phosphate. There is popularity known as phosphate-solubilising bacteria or fungi, generally abbreviated as PSB. A field study at IARI revealed that Mussoorie rock phosphate (MRP) was only 40.5% as effective as ordinary super phosphate but its efficiency increased to 79.7% when MRP was applied with PSB. The PSB also showed residual effects on succeeding maize equal to that obtained with ordinary super phosphate.

8. Efficient water management

The soil moisture influences the response to other inputs. The availability of fertilizer may not be seriously affected unless the available soil moisture is reduced to a very low level. However, the utilization of plant nutrients is commonly affected due to direct moisture effect on various plant growth processes. The input use efficiency is, therefore, high when adequate soil moisture is available. Under rainfed conditions the knowledge of available moisture in the soil profile is the most important factor for increasing fertilizer-use efficiency.

A field study at IARI indicated that recovery of fertilizers. N applied at 40 kg N/ha was very low (2-17%) when wheat was sown at a soil moisture tension of 1.5 bars or above without post-sowing irrigation. The apparent recovery increased as the pre-sowing soil moisture tension decreased. The post-sowing irrigation gave considerable increase in the recovery of applied nitrogen.

9. Weed control

Weeds take a heavy toll of plant nutrients. Data available from a number of studies clearly show that by effective control of weeds manually or through herbicides, the crop yield can be largely increased. Effective weed control can also increase the efficiency of fertilizer use and other inputs. For example, with the same amount of fertilizer rice yield in weed free plot was 5.2 tonnes/ha as against 2.7 in weedy plot. The corresponding agronomic efficiency (kg grain/kg fertilizer) was 23 in weed-free and 12 in weedy plots.

10. Crop Diversification

Crop diversification refers to bringing about a desirable change in the existing cropping patterns towards more balanced cropping systems to meet the ever-increasing demand for cereals, pulses, oilseeds, fibre, fodder, fuel etc. and aims to improving the soil health and agro-ecosystem. The need for diversification from the predominant rice-wheat system is seriously felt in Punjab due to marketing problem of both rice and wheat, excessive mining of groundwater, deterioration in soil health, multiplication of pests, diseases and weeds, intensive use of energy, reduced availability of protective food like pulses and oilseeds, and deterioration in the overall agro-ecosystem as a result of the dominating rice-wheat system.

Crop diversification could be undertaken both in *kharif* and *rabi* seasons. Maize, pigeon pea, soybean, groundnut and cotton could replace rice in upland welldrained soils. Similarly, in *rabi*, potato, mustard, chickpea, lentil can substitute wheat crop. Partial substitution of wheat is possible either by intercropping systems or by accommodating a third crop in the beginning or the later part of the crop growth period of which sowing one row of raya (*Brassica juncea*) alternating with 8 rows of wheat gives an additional yield of raya (0.23 tonne/ha) without affecting wheat yield. Growing a mungbean crop during summer and incorporation of its residue for green manuring increased the productivity of rice-wheat cropping by 1 tonne/ha besides producing 0.5 tonne/ha protein-rich pulse grain.

11. Integrated disease and pest management (IDPM)

Besides largescale depletion of essential plant nutrients by the high yielding varieties of cereals, growing of a particular crop over a long period of time, especially when large masses of land are under one variety can create ecological problem of high degree and may lead to epidemics of plant pathogens. For example, brown plant hopper (Nilaparvata lugens) was only a minor pest in pre-HYV era of rice, but today it is the most serious insect pest of rice. Among diseases bacterial leaf blight (Xanthomonas oryzae pv. oryzae), rice blast

(Magnaporthe grisea), sheath blight (Rhizoctonia solani), false smut (Ustilaginoidea virens) have caused serious damage in several countries of south-east Asia. To combat increasing disease and pest pressures in intensive cropping systems, there should be promotion and adaptation of IDPM, which involves resistant cultivars, biological control, crop rotation, appropriate agronomic practices and judicious use of pesticides/fungicides.

Indicators of sustainability, adaptation and mitigation

Evaluating alternative sustainable farming systems

A sustainable farming system is not simply a series of production practices that can each be evaluated independently from one another, but rather a series of production practices that are integrated and interrelated with each another. Sustainability is determined by the system as a whole, not its individual 322 components. He argues that what he calls "synergism" is the key to sustainability. The interdependent linkages between production practices comprising a farming system makes evaluation of the profitability and environmental benefits attributed to a specific practice difficult.

A specific production practice, say a particular type of tillage, when taken alone, may appear to be inconsistent with the goals of sustainable agriculture. However, it is the farming system in place, that is, the integrated system of interrelated production practices that ultimately determines the sustainability of the farming system.

Any sustainable farming system may include some specific production practices that might not appear to be consistent with the goals of sustainable agriculture, and yet, when integrated with other specific production practices, the entire farming system in total might be quite sustainable.

Identifying specific farming systems consistent with sustainable agriculture

The two underlying themes that appear in most definitions of sustainability and sustainable farming systems deal with (1) the economic profitability of the farming system over a long period of time; and (2) long-term benefits to the environment. Environmental benefits are sometimes not measured as an overall improvement in environmental quality over time, but instead compared with what would have happened to the environment over time had conventional (previously-employed, non-sustainable) production practices been continued. For example, it is generally recognized that any type of agricultural land use will result in a significant loss in top soil. Even idle land in grass steadily loses topsoil. An environmental goal of a sustainable farming system might not be to actually increase the quantity of topsoil on the farm, but rather to employ a farming system that minimizes the amount of topsoil loss over time, especially when compared with alternative farming systems that might instead have been continued. Thus, farming systems cannot simply be divided into two dichotomous categories, labelled either conventional or sustainable. Instead, there are degrees of sustainability. An existing farming system— one that might be described as conventional—may be profitable even over a long time period, consistent with one of the primary goals of sustainability. Furthermore, a farming system labelled as sustainable because the probable benefits to the environment over the long term are great may incorporate a number of specific production practices that, if taken individually, would be called conventional. A sustainable farming system does not necessarily employ an entirely different set of specific production practices and does not necessarily preclude the use of some specific practices that might be labelled "conventional."

Sustainable farming systems are, indeed, systems. In this context, a sustainable farming system must consist of a series of related and integrated production practices. In some instances, it may be possible to determine if a specific production practice incorporated into a farming system is more or less sustainable than another alternative production practice. For example, a specific production practice that makes better use of green manure crops than chemical fertilizers to improve soil fertility might result in environmental benefits arising from decreased ground and surface water contamination. Such a production might be labelled as sustainable based on perceived environmental benefits. Furthermore, differences in profitability that occur might be directly attributed to differences in the specific production practices that are employed. The farmer who reduces purchased chemical fertilizer use by relying more heavily on green manure crops in a rotation to improve soil fertility will likely experience some change in the pattern of profitability over time. Presumably, to the extent that profits change, the change occurred because of the modification in the specific production practice that was employed.

From the perspective of sustainability, an ideal situation would be one in which profits increase as a result of shift from chemical fertilizers to green manure in a rotation, and the environment is also significantly enhanced because of reduced nitrate pollutants in ground and surface water. Further, this could actually happen. A farmer who reduces chemical fertilizer use will likely decrease out-of-pocket expenses, enhancing profitability. But output levels may not remain constant, either, since profit is the net of revenue over costs. Moreover, the improvements to the quality of ground and surface water by reducing or eliminating chemical fertilizer use may not be significant or even measurable.

In most cases, however, the concept of a sustainable farming system suggests forgoing some profit (in comparison with the production system previously employed) over the short run (the first few years a farming system is in place) with the expectation that benefits will be achieved over a longer period of time. Long-run profitability may be increased relative to what would have occurred if the conventional farming system had continued indefinitely. While the ideal would be improved environmental quality over time, the sustainable farming system may be justified (and considered successful) if the benefit is only that the environment is less harmed than would have been the case if the conventional farming system had continued to be employed.

Thus, environmental benefits from alternative, sustainable farming systems must be evaluated not only in terms of absolute improvement in environmental quality, but also in relative terms, that is, relative to what would have occurred had the new, sustainable farming system not been implemented. Similarly, the consequences of such a sustainable farming system on profitability must be evaluated not only over a multi-year time horizon, but also relative the likely profitability of the conventional system over the same, multi-year time horizon.

Environmental considerations

The environmental benefits associated with sustainable farming systems can thus be categorized into four major groups:

1. Benefits accruing from a reduction in soil erosion due to wind and water
2. Benefits accruing from a reduction of pollutants in ground and surface water linked to chemical fertilizers primarily nitrates, but also phosphates
3. Benefits accruing from a reduction in pollutants in ground and surface water and in the air arising from herbicides and insecticides
4. A larger and more nebulous category of benefits that occur because, for example, soil structure might be maintained and enhanced with certain crop rotations, the use of animal manure, and other similar benefits arising from specific farming practices that help maintain and improve the productivity of the land over a long period of time. Heimlich argues that improvement of wildlife habitat should be an important goal. Among advocates, this category of benefits is quite important. For agricultural scientists, rationalizing sustainable farming systems based on these kinds of benefits is controversial. In many instances, the scientific evidence in support of these benefits is inclusive, or has not been conducted over a sufficient period of time such that the benefits, if any, can be measured.

It is tempting to define as sustainable only those farming systems that produce environmental gain. However, as earlier indicated, the diverse array of environmental benefits and damages makes it difficult to compress the various facets of environmental quality into a single measure or indicator. Each alternative farming system, whether labeled as sustainable or not, will generate a unique combination of environmental benefits and damages. A new farming system, for example, may significantly reduce soil erosion, but at the cost of additional ground water contamination relative to a farming system that had been previously employed. Questions arise that are not easily answered. Is such a farming system sustainable? Must all environmental consequences of a new farming system at least be no worse than what existed under the previously employed system?

Other questions pose additional difficulties. Are there tradeoffs between various categories of environmental benefits? If so, in valuing environmental benefits, what weights should be employed for each type or category of benefits? Should these weights be constant across states and regions? Is a ton of soil loss from erosion in an area where the topsoil is several feet thick as serious an environmental concern as a similar amount of loss from an area where the topsoil is fragile and only a few inches thick? Should greater weight be placed on reducing pollutants in instances where scientific evidence exists that a pollutant is harmful to human health, or should a reduction in any kind of pollutant be equally valued? These are difficult questions to answer.

Environmental benefits (and damages) can be categorized with respect to whether the benefits (and damages) occur on-site or off-site. A farmer who implements a production practice that reduces nitrate pollution in drinking water from a farm well is realizing an on-site (benefit to the farmer) environmental benefit, whereas, if the production practice reduces nitrate contamination in wells of neighboring farms, an off-site benefit (benefit to others) occurs. If additional costs (and perhaps a reduction in profitability) are incurred from a particular production practice that also provides environmental benefits (or reduces harm to the environment), farmers would likely be more interested in implementing practices that provide primarily on-site benefits (benefits to them) than primarily off-site benefits (benefits to others).

A farming system that reduces soil erosion from water provides long-term on-site benefits to the farmer in the form of a reduced rate of loss of soil productivity over time. However, the reduction in silting of rivers arising from reduced water erosion may be highly beneficial to others, including taxpayers who must pay for the cost of dredging silted rivers. In this instance, the private interests of the farmer and the public interest of others coincide. In general, sustainable farming systems that reduce soil erosion provide considerable private on-site benefits.

The public off-site benefits may be noticeable, however, only if a comparatively large number of farmers adopt production practices that lead to a significant reduction in soil erosion in an area.

Aside from the water well example, the on-site benefits to farmers of reducing the use of chemical fertilizers and pesticides may be somewhat less clear. Some farmers and soil scientists have argued that monocultures employing chemicals ultimately lead to a deterioration of the soil structure over time, with consequent negative effects on the long-term productivity of land. The long-term safety of certain agricultural chemicals to farmers their families and hired employees is another concern. Agricultural scientists who deal with pesticides are equally convinced of the current safety of the products, if applied in the manner and in the amounts as labeled.

6

Conservation Agriculture Strategies in Agriculture

Conservation agriculture (CA) defined as minimal soil disturbance (no-till) and permanent soil cover (mulch) combined with rotations is a recent agricultural management system that is gaining popularity in many parts of the world. Cultivation is defined by the Oxford English dictionary as "the tilling of land", "the raising of a crop by tillage" or "to loosen or break up soil". Other terms used in this dictionary include "improvement or increase in (soil) fertility". All these definitions indicate that cultivation is synonymous with tillage or ploughing.

The other important definition that has been debated and defined in many papers is the word "sustainable". The Oxford dictionary defines this term as "capable of being borne or endured, upheld, defended, maintainable". Something that is sustained is "kept up without intermission or flagging, maintained over a long period". This is an important concept in today's agriculture since the human race will not want to compromise the ability of its future offspring to produce their food needs by damaging the natural resources used to feed the population today.

Issues related to tillage

Cultivation techniques or tillage

The following summarizes the reasons for using tillage:

1. Tillage was used to soften the soil and prepare a seedbed that allowed seed to be placed easily at a suitable depth into soil moisture using seed drills or manual equipment. This results in good, uniform seed germination.
2. Wherever crops grow, weeds also grow and compete for light, water and nutrients. Every gram of resource used by the weed is one less gram for the crop. By tilling their fields farmers were able to shift the advantage from the weed to the crop and allow the crop to grow without competition early in its growth cycle with resulting higher yield.
3. Tillage helped release soil nutrients needed for crop growth through mineralization and oxidation after exposure of soil organic matter to air.

4. Previous crop residues were incorporated along with any soil amendments (fertilizers, organic or inorganic) into the soil. Crop residues, especially loose residues, create problems for seeding equipment by raking and clogging.
5. Many soil amendments and their nutrients are more available to roots if they incorporated into the soil; some nitrogenous fertilizers are also lost to the atmosphere if not incorporated.
6. Tillage gave temporary relief from compaction by using implements that could shatter below ground compaction layers formed in the soil.
7. Tillage was determined to be a critical management practice for controlling soil borne diseases and some insects.

There is no doubt that this list of tillage benefits was beneficial to farmers, but at a cost to him and the environment and the natural resource base on which farming depended. The utility of ploughing was first questioned by an agronomist in the 1930s. The tragic dust storms in the mid-western United States in the 1930s was a wakeup call to how man's interventions in soil management and ploughing led to un-sustainable agricultural systems.

Conservation tillage and conservation agriculture

Since the 1930s and during the next 75 years members of the farming community have been advocating a move to reduced tillage systems that use less fossil fuel, reduce runoff and erosion of soils and reverse the loss of soil organic matter. The first 50 years was the start of the conservation tillage movement and today a large percentage of agricultural land is cropped using these principles. However, in the book "No-Tillage Seeding" (Baker *et al.*, 2002) explains "As soon as the modern concept of reduced tillage was recognized, everyone, it seems, invented a new name to describe the process." He defines conservation tillage as:

"Conservation tillage is the collective umbrella term commonly given to no-tillage, direct-drilling, minimum-tillage and/or ridge-tillage, to denote that the specific practice has a conservation goal of some nature. Usually, the retention of 30% surface cover by residues characterizes the lower limit of classification for conservation-tillage, but other conservation objectives for the practice include conservation of time, fuel, earthworms, soil water, soil structure and nutrients. Thus residue levels alone do not adequately describe all conservation tillage practices" (Baker *et al.*, 2002).

FAO has characterized conservation agriculture as follows:

"Conservation Agriculture maintains a permanent or semi-permanent organic soil cover. This can be a growing crop or dead mulch. Its function is to protect the soil physically from sun, rain and wind and to feed soil biota. The soil micro-organisms and soil fauna take over the tillage function and soil nutrient balancing.

Mechanical tillage disturbs this proccss. Therefore, zero or minimum tillage and direct seeding are important elements of CA. A varied crop rotation is also important to avoid disease and pest problems." (FAO web site).This has led to confusion among the agricultural scientists and more important the farming community. To add to the confusion, the term "conservation agriculture" has recently been introduced by FAO (Food and Agriculture Organization web site) and others and its goals defined by FAO as follows:

"Conservation agriculture (CA) aims to conserve, improve and make more efficient use of natural resources through integrated management of available soil, water and biological resources combined with external inputs. It contributes to environmental conservation as well as to enhanced and sustained agricultural production. It can also be referred to as resource efficient or resource effective agriculture" (FAO)

Table. Extent of Adoption of CA Worldwide 2013 updates

Country	CA area '000 ha
USA	35,613.00
Brazil	31,811.00
Argentina	29,181.00
Canada	18,313.00
Australia	17,695.00
China	6.670.00
Russia	4,500.00
Paraguay	3,000.00
Kazakhstan	2,000.00
India	1,500.00
Uruguay	1,072.00
Spain	792.00
Bolivia	706.00
Ukraine	700.00
Italy	380.00
South Africa	368.00
Zimbabwe	332.00
Venezuela	300.00
Finland	200.00
France	200.00
Zambia	200.00
Germany	200.00
Chile	180.00
New Zealand	162.00
Mozambique	152.00
United Kingdom	150.00

Advantages of Conservation Agriculture

Conservation agriculture is generally a "win-win" situation for both farmers and the environment. Yet many people intimately involved with worldwide food production have been slow to recognize its many advantages and consider it to be a viable alternative to conventional agricultural practices that are having obvious negative impact on the environment. Much of this has to do with the fact that conservation agriculture requires a new way of thinking about agricultural production in order to understand how one could possibly attain higher yields with less labor, less water and fewer chemical inputs. In spite of these challenges, conservation agriculture is spreading to farmers throughout the world as its benefits become more widely recognized by farmers, researchers, scientists and extension workers alike.

Specifically, conservation agriculture (CA) increases the productivity of:

- **Land** - Conservation agriculture improves soil structure and protects the soil against erosion and nutrient losses by maintaining a permanent soil cover and minimizing soil disturbance. Furthermore, CA practices enhance soil organic matter (SOM) levels and nutrient availability by utilizing the previous crop residues or growing green manure/ cover crops (GMCC's) and keeping these residues as a surface mulch rather than burning. Thus, arable land under CA is more productive for much longer periods of time.

- **Labor** - Because land under no-till is not cleared before planting and involves less weeding and pest problems following the establishment of permanent soil cover/crop rotations, farmers in Ghana reported a 22% savings in labor associated with maize production. Similar reductions in labor requirements have been reported with no-till rice-wheat systems in South Asia and various CA technologies in South America. Much of the reduced labor comes from the absence of tillage operations under CA, which use up valuable labor days during the planting season.

- **Water** - Conservation agriculture requires significantly less water use due to increased infiltration and enhanced water holding capacity from crop residues left on the soil surface. Mulches also protect the soil surface from extreme temperatures and greatly reduce surface evaporation, which is particularly important in tropical and sub-tropical climates. In Sub-Saharan Africa, as with other dryland regions, the benefits of conservation agriculture are most salient during drought years, when the risk of total crop failure is significantly reduced due to enhanced water use efficiency.

- **Nutrients** - Soil nutrient supplies and cycling are enhanced by the biochemical decomposition of organic crop residues at the soil surface that are also vital for feeding the soil microbes. While much of the nitrogen needs of primary food crops can be achieved by planting nitrogen-fixing legume species, other plant essential nutrients often must be supplemented by additional chemical and/or organic fertilizer inputs. In general, soil fertility is built up over time under conservation agriculture, and fewer fertilizer amendments are required to achieve optimal yields over time.

- **Soil biota** - Insect pests and other disease causing organisms are held in check by an abundant and diverse community of beneficial soil organisms, including predatory wasps, spiders, nematodes, springtails, mites and beneficial bacteria and fungi, among other species. Furthermore, the burrowing activity of earthworms and other fauna create tiny channels or pores in the soil that facilitate the exchange of water and gases and loosen the soil for enhanced root penetration.
- **Economic benefits** - Farmers using CA technologies typically report higher yields (up to 45-48% higher) with fewer water, fertilizer and labor inputs, thereby resulting in higher overall farm profits. In Paraguay, net farm income of no-till (NT) farming on large-scale commercial farms increased from $2,3467 to $32,608 more than farms using conventional tillage over a 10 year period. The economic benefits of NT and other conservation agriculture technologies, more than any other factor, have lead to widespread adoption among both large- and small-scale farmers throughout the world.
- **Environmental benefits** - Conservation agriculture represents an environmentally-friendly set of technologies. Because it uses resources more efficiently than conventional agriculture, these resources become available for other uses, including conserving them for future generations. The significant reduction in fossil fuel use under no-till agriculture results in fewer greenhouse gases being emitted into the atmosphere and cleaner air in general. Reduced applications of agrochemicals under CA also significantly lessen pollution levels in air, soil and water.
- **Equity considerations** -Conservation agriculture also has the benefit of being accessible to many small-scale farmers who need to obtain the highest possible yields with limited land area and inputs. Perhaps the biggest obstacle thus far for the technology spreading to more small-scale farmers worldwide has been limited access in certain areas to certain specialized equipment and machinery, such as no-till planters. This problem can be remedied by available service providers renting equipment or undertaking conservation agriculture operations for farmers who would not otherwise have access to the needed equipment. Formulating policies that promote adoption of CA are also needed. As more and more small-farmers gain access to CA technologies, the system becomes much more "scale neutral."
- **Active role for farmers** -As with any new agricultural technology, CA methods are most effective when used with skillful management and careful consideration of the many agro-ecological factors affecting production on any given farm or field. Rather than being a fixed technology to be adopted in blueprint-like fashion, CA should be seen as a set of sound agricultural principles and practices that can be applied either individually or together, based on resource availability and

other factors. For this reason, farmers are encouraged to experiment with the methods and to evaluate the results for themselves- not just to "adopt" CA technologies. Selecting among different cover crop species, for example, needs to be determined in relation to particular agroecological conditions of the farm, including soil type, climate, topography as well as seed availability and what the primary function of the GMCC will be. Similarly, planting distances, irrigation requirements and the use of agrochemicals to control weeds and pests among other considerations, must be decided based on what the farmer needs as well as the availability of these and other resources.

Table 2: A comparison of traditional tillage, conservation tillage (CT) and conservation agriculture (CA) for various issues.

Issues	Traditional tillage (TT)	Conservation tillage (CT)	Conservation agriculture (CA)
Practice	Disturbs the soil and leaves a bare surface	Reduces the soil disturbance in TT and keeps the soil covered	Minimal soil disturbance and soil surface permanently covered
Erosion	Wind and soil erosion maximum	Wind and soil erosion reduced significantly	Wind and soil erosion the least of the three
Soil physical health	The lowest of the 3	Significantly improved	The best practice of the 3
Compaction	Used to reduce compaction but can also induce it by destroying biological pores.	Reduced tillage is used to reduce compaction	Compaction can be a problem but use of mulch and promotion of biological tillage helps reduce this problem.
Soil biological health	The lowest of the 3 because of frequent disturbance	Moderately better soil biological health	More diverse and healthy biological properties and populations
Water infiltration	Lowest after soil pores clogged	Good water infiltration	Best water infiltration
Soil organic matter	Oxidises soil organic matter and causes its loss	Soil organic buildup possible in the surface layers	Soil organic buildup in the surface layers even better than CT.
Weeds	Controls weeds but also causes more weed seed to germinate	Reduced tillage controls weeds but also exposes other weed seed for germination	Weeds are a problem especially in the early stages of adoption, but problems are reduced with time and residues can help suppress weed growth
Soil temperature	Surface soil temperature more variable	Surface soil temperature intermediate in variability	Surface soil temperature moderated the most.
Diesel use and costs	Diesel use high	Diesel use intermediate	Diesel use much reduced
Costs production	Highest costs	Intermediate costs	Lowest costs
Timeliness	Operations can be delayed	Intermediate timeliness of operations	Timeliness of operations more optimal
Yield	Can be lower where planting delayed	Yields same as TT	Yields same as TT but can be higher if planting done more timely

From Soil Conservation to Conservation Agriculture

Why Soil Conservation?

Erosion

Agriculture is world-wide causing serious soil losses. If the destruction of agricultural soils continues in the same way, humans might face serious problems to feed a growing population. There are different causes for this inadequate use of the soil. In many developing countries the hunger is forcing poor people to cultivate areas which are not suitable for agricultural use and which only with major and costly efforts, like the construction of terraces, can be sustainably converted in agricultural land.

However, serious damage, because it is on large scale, is also caused by mechanised farming. As an example may serve the dust bowls of the 30s in the USA that destroyed vast areas of fertile land through wind erosion. The same mistakes are at present still causing important soil losses in the agriculture world-wide.

Erosion has become a direct threat to farmers. Systems where developed to control erosion and conserve the soil which means to avoid that the soil moves from one place to another. Evidence of these concepts for the control of water erosion are contour planting, contour bunds and trenches to avoid down-slope water runoff. In other places, great effort was spent in the construction of terraces. It was recommended not to leave the bare soil unprotected but to cover the surface with stubble or other mulch to break the kinetic energy of wind and water. In short, many efforts where undertaken to avoid that wind and water would move the soil.

Water Conservation

With all these measures it was not reflected that erosion is not the main problem but only a consequence of the way agricultural soils are treated, particularly in mechanized agriculture. As example might serve the western plains of Nicaragua. This area with the most fertile soils of the country has always been intensively cultivated. Over the last 40 years it converted to a cotton growing area. With the cultivation which until present is nearly exclusively done by disc harrows the soil erosion problems increased. As a solution to the problem terraces were constructed following strictly the contour lines. The result were irregularly shaped plots some of which could hardly been cultivated by a tractor. Those terraces were tilled over 20 or 30 years always with disc tools without any change in implements. To make it worse, tractors had on the small plots to carry out frequent turns. At present, all the soil in the western plains of Nicaragua has serious compaction problems. But what is even worse, the compaction does not permit water infiltration (Kayombo and Lal, 1994). To avoid water logging on the terraces the water is led away in drainage canals. As a consequence of this, the area is now characterized by enormous erosion gullies and the ground water table has considerably fallen.

This shows that soil loss through erosion is only part of the problem. The loss of rain water that cannot infiltrate in the soils to replenish the ground water reserves might on the long term be the more serious problem.

Consequently, the way soil is cultivated must be drastically changed. Soil erosion and water loss is not controlled by mechanical means but only by a living and stable soil structure. Only this can avoid that water runs on the surface rather than being absorbed as completely as possible by the soil.

The Concept of Integrated Soil Management - The Conservation Agriculture

Soil Tillage Concepts within Conservation Agriculture

Unfortunately there exists no mechanical implement that could create a stable soil structure. Mechanized soil tillage can only destroy this structure. Therefore we need a different concept of soil tillage and profound knowledge of the type of intervention each implement is carrying out in the soil.

Certainly it must be differentiated between different soils and their susceptibility to soil structure losses. However, in any case a stable and optimal soil structure for plant growth as well as for water infiltration and erosion control is only achieved by living biological processes in the edaphon like for example the creation of humus.

When to Till the Soil

Under the above concept the best form of mechanized soil tillage is not to do any. However, the concepts of zero-tillage are not applicable in all cases. Agriculture is always an artificial intervention in natural processes and therefore it has to be accepted that, from case to case, corrective interventions have to be done. Even under zero tillage concepts, some sort of tillage is done through traffic for planting, pest control and harvest. Traffic causes compaction and is as such one form of tillage.

Each time a problem occurs that might call for a tillage intervention the problem should be carefully analyzed to find out a way to control it with a minimum intervention in the soil.

Within possible interventions 5 different basic operations of soil tillage can be distinguished:

- turning
- mixing
- loosening
- pulverizing
- compacting

In addition to those basic operations of soil tillage some other agricultural operations have a direct effect on the soil:

- mechanical weed control
- shaping of surfaces - ridging, leveling
- harvesting crops like potatoes, beets, peanuts

Each tillage implement carries out a specific spectrum of those basic operations. The knowledge of these characteristics and the availability of adequate equipment allow to limit the intervention to the minimum necessary. Some of the operations of the second category cannot be avoided. But the majority of the basic operations of the first group are not essential for agriculture. This accounts particularly for the turning operation which represents the most drastic intervention in the soil.

Turning

The most suitable implement for this operation is the mouldboard plough. However, the necessity to bury surface material into the soil and bring soil of deeper horizons to the surface is very limited. The argument that ploughing controls weeds is not necessarily true if the operation is carried out annually: in this case seeds from the last year are brought back to the surface while fresh seeds are conserved for the next year. The use of the plough originated and was justified in situations of limited traction power and with simple planting equipment that required a clean soil surface for proper seeding.

Mixing

This operation can be done with implements like a chisel plough or heavy tine cultivator. It could be justified to facilitate the decomposition of stubble or surface mulch. The depth of intensive mixing is usually not more than 10 cm.

Loosening

This operation is best carried out with a Paraplow which allows loosening without any other intervention in the soil. Under situations of a compacted soil or a soil with an unstable structure this operation creates sufficient pores in the soil to permit water infiltration. However the residual effect of this treatment varies a lot depending on the soil characteristics and subsequent operations (Kayombo and Lal, 1994).

Pulverising

This operation was formerly required for seedbed preparation. For that purpose only a very shallow superficial horizon was necessary to be pulverized. The pulverization of deeper horizons, as it can be achieved with disc harrows or rotavators, is in no case justified. At present, the technology for planting the majority of field crops without the need of a seedbed exists. Only in very few cases, mainly in horticulture, the fine seedbed preparation might still be required.

Compacting

This operation becomes necessary after a deep loosening operation has been carried out shortly before planting. Compacting is required to assure the capillary contact to the soil and groundwater. For the same reason the seed is after planting usually pressed into the soil which represents also a small compacting operation.

The necessity of many of the above operations resulted from the deficiencies of planting equipment. Definitively none of the operations as such can create an ideal soil structure.

With the disc harrow all five operations are carried out at the same time: the soil is turned, although not as completely as with the moldboard, it is mixed, loosened, pulverized over the entire working depth by the friction on the rotating discs and compacted beneath the cutting edges of the disc. The long term result is a degraded soil with a noticeable compaction horizon.

How to practice conservation agriculture

Soil management in Conservation Agriculture

Each context brings different problems regarding the implementation of Conservation Agriculture, and many technologies can be adapted, including traditional ones. Technical solutions should be found through close partnerships between farmers, private sector industries, extension services and researchers. This approach has already permitted to develop and adapt Conservation Agriculture for small and big farms, manual, animal traction and mechanized agriculture, in various agro-ecological zones and socio-economic contexts, and for many farming systems

The physical and chemical structure, and biological activity of soil, including the spatial arrangement of soil components, are fundamental to sustaining agricultural productivity and determine in their complexity the soil health and fertility. Soil management should maintain and improve soil fertility by minimizing losses of soil, nutrients, and agrochemicals from erosion, runoff, and leaching into surface- or ground water, and surrounding natural vegetation and wildlife in absolute terms as well as in relative terms below the self recovery of the respective ecosystems. Good agricultural practice should:

- Establish a detailed knowledge of the nature, properties, distribution, and potential uses of soils of the farm.
- Avoid mechanical soil disturbance to the extent possible.
- Avoid soil compaction beyond the elasticity of the soil.
- Maintain or improve soil organic matter during rotations until reaching an equilibrium level.
- Maintain organic cover through crop residues and cover crops to minimize erosion loss by wind and/or water.
- Maintain balanced nutrient levels in soils.
- Avoid contamination with agrochemicals, organic and inorganic fertilizers and other contaminants by adapting quantities, application methods and timing to the agronomic and environmental requirements.
- Maintain a record of the annual use and inputs and outputs of each individual land-management unit.

Maintaining and managing a permanent soil cover

Farm planning and crop rotations design

The planning of farm activities in Conservation Agriculture systems requires skills, knowledge and moreover, practice. It has to take in account crop rotations (2-3 year planning), soil cover management, other farm activities such as livestock, marketing or food processing as well as other factors (climate, labour availability, etc.). The design of crop rotations and the choice and management of cover crops must ensure that the biomass production is sufficient to satisfy all the needs (permanent soil cover, human and livestock feeding, fibres, etc.), that soil water and nutrient resources are adequate for the crop and that the use of agrochemical inputs such as desiccants and herbicides can be minimized.

Choice of cover crops

Cover crops must be multi-purpose crops and must be carefully managed through appropriate technologies. For example Crotalaria juncea L. is a good source of fibre, very effective for nematode control as well as soil properties and fertility improvement (N fixation, high level of Organic matter, etc.). This cover crop is used in orchards or in rotation with many crops. This choice of cover crops must also include socio-economic criteria such as access to appropriate seeds and possible seed production by the farmer.

Management of crop residues and cover crops

After harvest, crop residues are left in the field, usually without chopping them. Cover crops may be managed by using desiccants or by mechanical means such as knife rollers – recommended for small farmers. Mechanical tools have been developed by manufacturers, but many of them can be adapted by farmers themselves. However, they must be adapted to the cover crop (appropriate weight) and must not harm the soil.

Minimum soil disturbance – zero tillage and direct planting

Planting crops through the soil cover can be done by direct seeding, direct planting or broadcasting into the soil cover, depending on the specific conditions (soil, climate, seeds and cover properties). Most important before introducing zero tillage is to correct soil limitations such as hard pan, levelling, surface crusting or compaction, and eventually to address chemical limitations by incorporating lime, organic matter or other amendments. However, the introduction of appropriate crops in the rotation – selected for the properties of their roots, for example use of radish for decompaction – must ensure this role from the first year of introduction.

Most machinery already exists for manual, animal-drawn or mechanised agriculture. However, if it is not locally available or accessible through import, it is an option for example, farmers, researchers and local blacksmith (or other professionals) to develop the appropriate tools. In this task, they may be supported by experts or companies from other countries or regions in order to benefit from existing experiences.

Pest, weed and fertility management

Pest, weed and fertility management are critical issues in any farming system. Before implementing Conservation Agriculture, any limitation related to pest, weed and soil fertility must be assessed and addressed. This may imply chemical and labour inputs at the beginning which might partly exceed conventional levels. To the extent that a new balance between the organisms and the farm ecosystems – pests and beneficial organisms, crops and weeds – becomes established and the farmer learns to manage the cropping systems, the use of synthetic pesticides, herbicides and mineral fertilizer tends to decline to a level below the original "conventional" farming.

In case of a lack of resources or access to chemical inputs, some solutions have been found by farmers: weed control with mechanical tools and through crop rotations; use of manure, Biological nitrogen fixation (BNF) and other crops for soil fertility; home-made "soups" for disease control, etc. Combining Conservation Agriculture with Organic Agriculture is also an option which has been adopted by mostly small scale farmers.

Pest management

New, or in their appearance so far unknown pests, might occur in the beginning, which may involve inputs of chemicals. However, crop rotations and appropriate use of cover crops should also help in this task, since the beginning, by interrupting the infection chain between subsequent crops. In Conservation Agriculture systems, pest and disease control are based on Integrated Pest Management (IPM) technologies. Again, information, knowledge and training are very important in IPM, and Farmer Field Schools (FFS) one possible approach to enhance farmer's understanding of biological processes and their capabilities to respond to new problems.

Weed management

Synthetic chemical herbicides may be necessary in the first years, but have to be used with very much care to reduce the negative impact on soil life. Mulch cover and cover crops, crop rotations, desiccants and herbicides are the technologies to control weeds. Mechanical weed control through rolling, cutting or breaking can also be applied. For exceptional cases mechanical weeders exist for animal traction and tractors which can work through a mulch cover. With time, the amount of herbicides needed decreases because the farmer becomes more experiences in managing the system and the weed population decreases. Biological weed control may be very effective through appropriate choice of cover crops. In southern Brazil, for example, Black Oat is integrated in the rotations to suppress weeds. It can also be used as fodder and its straw is quite persistent, which is an advantage in areas where crop residues tend to mineralise too quickly to ensure surface cover.

Fertility management

Any soil limitations, such as acidity, salinity or toxicity problems must be addressed before and during the implementation of Conservation Agriculture. Soil Organic Matter (SOM) and judicious choice of crops and cover crops play a key role in soil fertility management, and especially SOM provided by root decomposition. The integration of livestock in the system can also be a solution, especially for the farmers with poor resources or access to mineral fertilisers. The use of organic and mineral fertilisers may be necessary. Appropriate application can be done with combined tools for direct seeding-fertilisation.

Conservation agriculture – The way forward

Conservation agriculture has emerged as a new paradigm to achieve goals of sustainable agricultural production. It is a major step towards transition to sustainable agri-culture. The term CA refers to the system of raising crops without tilling the soil while retaining crop residues on the soil surface. The key elements which characterize CA include:

- Minimum soil disturbance by adopting no-tillage and minimum traffic for agricultural operations,
- Leave and manage the crop residues on the soil surface, and
- Adopt spatial and temporal crop sequencing/crop rotations to derive maximum benefits from inputs and minimize adverse environmental impacts.

Combining the above elements with improved land-shaping (e.g. through laser aided levelling, planting crops on beds, etc.) further enhances the opportunities for improved re-source management. In conventional systems, while soil tillage is a necessary requirement to produce a crop, tillage does not form a part of this strategy in CA. Intensive tillage in conventional systems causes gradual decline in soil organic matter content through accelerated oxidation, resulting in reduced capacity of the soil to regulate water and nutrient supplies to plants. Burning of crop residues, a common practice in many areas (e.g. rice–wheat cropping system) further causes pollution, GHG emission and loss of valuable plant nutrients. When crop residues are retained on the soil surface in combination with no tillage, it initiates processes that lead to improved soil quality and overall resource enhancement.

Benefits of CA are several fold

Direct benefits to farmers include reduced cost of cultivation through savings in labor, time and farm power, and improved use efficiency resulting in reduced use of inputs. More importantly, CA practices reduce resource degradation. Gradual decomposition of surface residues improves soil organic matter status, biological activity and diversity and contributes to overall improvement in soil quality. CA is a way to reverse the processes of degradation inherent in conventional agricultural practices involving intensive cultivation, burning and/or removal of crop residues, etc. CA leads to sustainable improvements in efficient use of water and nutrients by improving nutrient balance and availability, infiltration and retention by the soil, reducing water loss due to evaporation and improving the quality and availability of ground and surface water.

Conservation agriculture in India

In India, efforts to adopt and promote resource conservation technologies have been underway for nearly a decade, but it is only in the past 4–5 years that technologies are finding acceptance by the farmers. This effort has been spearheaded by Rice–Wheat Consortium for Indo-Gangetic Plains, a CGIAR eco-regional initiative involving several CG centres and the National Agricultural Research Systems of India, Pakistan, Bangladesh and Nepal. Concerns about stagnating productivity, increasing production costs, declining resource quality, declining water tables and increasing environmental problems are the major forcing factors to look for alternative technologies, particularly in the north-west region encompassing Punjab, Haryana and western Uttar Pradesh (UP). In the eastern region covering eastern UP, Bihar and West Bengal, developing and promoting strategies to overcome constraints for continued low cropping system productivity have been the chief concern. The primary focus of developing and promoting CA practices has been the development and adoption of zero tillage cum fertilizer drill for sowing wheat crop in rice–wheat system. Other interventions being tested and promoted include raised-bed planting system, laser-aided land-levelling equipment, residue management alternatives, alternatives to rice–wheat cropping system in relation to CA technologies, etc. The area planted with wheat adopting zero-tillage drill has been rapidly increasing. It is speculated that over the past few years, adoption of zero-tillage has expanded to cover about 1 m ha. The rapid adoption and spread of zero tillage is attributed to benefits resulting from reduction in cost production, reduced incidence of weeds and therefore savings on account of weedicide costs, savings in water and nutrients and environmental benefits. Adopting CA systems further offers opportunities for achieving greater crop diversification. Crop sequences/rotations and agroforesting systems, when adopted in appropriate spatial and temporal patterns, can further enhance natural ecological processes which contribute to system resilience and reduced vulnerability to yield, thus reducing disease and pest problems. Zero-tillage when combined with appropriate surface-managed crop residues sets in processes whereby slow decomposition of residues results in structural improvement of soil and increased recycling and availability of plant nutrients. Surface residues are also expected to improve soil moisture regime, improve biological activity and provide a more favourable environment for growth. These processes, however, are slow and results are expected only with time.

In India, CA is a new concept and its roots are only now beginning to find ground. Globally, CA is being considered a route to sustainable agriculture and offers opportunities for moving to the next phase in Indian agriculture.

Conclusions

Crop production in the next decade will have to produce more food from less land by making more efficient use of natural resources and with minimal impact on the environment. Only by doing this will food production keep pace with demand and the productivity of land preserved for future generations. This will be a tall order for agricultural scientists, extension personnel and farmers. Use of productive but more sustainable management practices described in this book can help resolve this problem. Crop and soil management systems that help improve soil health parameters (physical, biological and chemical) and reduce farmer costs are essential. Development of appropriate equipment to allow these systems to be successfully adopted by farmers is a pre-requisite for success. Overcoming traditional mindsets about tillage by promoting farmer experimentation with this technology in a participatory way will help accelerate adoption. Encouraging donors to support this long term, applied research with sustainable funding is also an urgent requirement.

7

LEIA, HEIA and Its Techniques for Sustainability

Low External Input Agriculture (LEIA)

It is a form of agriculture wherein the use of external inputs including chemical inputs is minimized and production is kept sustainable. It uses synthetic fertilizers or chemical pesticides below rates commonly recommended by the Extension workers. There is no elimination of these materials. Any system that reduces purchased chemical inputs can be called low-input farming. Yields are maintained through greater emphasis on crop rotation, crop residues, animal manures, bio-fertilizers, IPM, IWM and utilization of on-farm resources and management. Locally available resources are used by maximizing the complementary and synergistic effects of different components of the farming systems. There is use of external inputs in a complementary way. Main advantages of this type of agriculture are:

1. Ecological balance
2. Low cost of production
3. Healthy environment
4. Overall risk of the farmer is considerably reduced
5. Ensure both short and long term profitability
6. Food very little or no pesticide residue is ensured
7. High diversity, renewable and biodegradable inputs are used
8. Rate of extraction of resources do not exceed rate of regeneration

Some of the disadvantages may be:

1. Lower yield
2. Ineffective control of weeds, insect-pests and diseases.

IPM is probably the oldest and widely accepted Extension Service programme devoted to low-input agriculture. However, only recently have the "non-chemical" approaches-such as cultural, mechanical, and biological-within the IPM framework been emphasized over the chemical component. Such programmes are now called biologically-intensive IPM.

LEIA systems are mainly based on preventive approach whereby the problem is tackled at its roots, as compared to symptom-curing nature of HEIA. Ecological and biological principles are the basis of the farm system. Nature works for the farmer, the farmer does not have to work against it. They are often based on ITK and production systems adapted to modern requirements and much less on external expertise.

- Supports profitable production
- Protects environmental quality
- Uses natural resources efficiently
- Provides consumers with affordable, high-quality products
- Decreases dependency on non renewable resources
- Enhances the quality of life for farmers and rural communities and
- Will last for generations to come.

Criteria for LEISA

Ecological criteria

1. Balanced application of plant nutrients
2. Judicious use of water
3. Diversity of genetic resources
4. Efficient utilization of genetic resources
5. Efficient utilization of energy sources
6. Reducing harmful impact on environment
7. Reducing dependence on external inputs

Economic criteria

1. Sustained farmer livelihood systems
2. Competitiveness
3. Efficient utilization of production factors
4. Reduced relative value of external inputs

Social criteria

1. Wide-spread and equitable adoption potential, especially among small farmers
2. Reduced dependency on external institutions
3. Enhanced food security at the family and national level

4. Respecting and building on indigenous knowledge, beliefs and value systems
5. Contribution to employment generation.

LEIA is a form of agriculture that maintains productivity for a long period by:

- Increasing the use of locally available resources which complement each other and thus have synergistic effects
- Reducing the use of off-farm external and non-renewable inputs with potential of degrading environment
- Improving the match between cropping patterns, productive potential and environmental constraints for longer sustainability of the system
- Making optimum use of biological and genetic potential of plant and animal species
- Taking full advantage of ITK and practices, innovative approaches not fully understood by scientists.

High External Input Agriculture (HEIA)

The population of developing world is increasing rapidly and the demand for food is not always keeping pace with population growth. The contribution of developing countries to world agricultural production is much lower than developed countries. Only those countries, which can maintain pace with the increasing population by way of increased production will be successful to avoid mass hunger. Before independence, Indian agriculture was considered as a gamble with monsoons. For the last few decades, there is remarkable progress in Indian agriculture inspite of decreasing per capita agricultural land availability. The Ford Foundation and the Rockefeller Foundation in the US encouraged the adoption of new technology (biological breakthrough & adoption breakthrough) and introduced in 1966 – *The Green Revolution*. This term was coined by William S. Gaud, Director USAID. In the last 3-4 decades cereal yields have doubled and have kept pace feed expanding population of our country. The increase in food grain production was achieved through the use of high yielding crop varieties, irrigation water and higher levels of inputs of fertilizers and plant protection chemicals. High amount of inputs were applied from external sources for high productivity. Consumption of chemical fertilizers has gone up seven times in the last 20 years, but production has only increased a miserable two-fold.

HEIA: It is a form of agriculture wherein use of external inputs is high for getting higher productivity of crops. Synthetic fertilizers are used to supply nutrients to crop plants and chemical pesticides are used to control pests. High analysis and low diversity chemical inputs are used. Rate of extraction of resources exceeds the rate of regeneration. Food products may contain residues of toxic chemicals. There are excessive tillage operations, excessive use of irrigation water, use of inorganic fertilizers, use of chemical pesticides, neglect of natural allies of farming and use of varieties inconsistent with local factors. But, high productivity resulted in degradation of natural resources i.e. soil, water resources and genetic diversity and adversely affected the quality of living environment by polluting soil, atmosphere, water and foodstuffs. Its sustainability at higher levels is possible only by adopting proper use of factors to help in maintaining natural resources.

High External Input Agriculture (HEIA) had following main impacts:

- Increase in production and productivity of crops
- Increase in cropped area
- Increased use of irrigation water
- Increase in consumption of fertilizers and pesticides
- Increase in market, storage and transportation facilities
- Improvement in farm credit facilities – establishment of rural credit societies
- Strengthening of research and extension facilities
- Consolidation of land holdings
- Exploitation of natural resources of land and water

Various ill effects of HEIA are as under:

- Depletion of soil fertility
- Indiscriminate killing of friendly insects, predators and microbes
- Development of resistant species of insect-pests
- Narrowing of genetic base of crop plants
- Deteriorating soil structure
- Poor aeration and water holding capacity
- Prone to soil erosion by wind and water
- Increase in demand of irrigation water for crops
- Development of soil salinity and poor drainage with big irrigation projects
- Depletion of ground water table
- Poor drought tolerance of crops
- Shift in weed flora and development of resistant biotypes of weeds
- Reduction in returns on inputs
- Environmental pollution with the use of agrochemicals and by their production units
- Harmful effect on the health of workers of production units of agro-chemicals and the farmers who use agro-chemicals

- Contamination of food material with pesticide residues
- Replacement of nutritious food crops with cash crops
- Increased socio-economic disparities
- Inflationary spirals because of increased input subsidies
- Increasing corruption in political and bureaucratic circles
- Destruction of local culture (commercialisation and consumerization)
- Leading financial institutions into disarray because of write-off of loans
- Sparking off social and political turmoil resulting in violence because of agriculture and economic problems
- Increased cost of crop production with the use of high inputs
- Depletion of fossil fuel resources.

8

Integrated Farming Systems-Historical Background, Objectives & Characteristics

What is a system?

A group of interacting components, operating together for a common purpose, capable of reacting as a whole to external stimuli is called a system. It is not affected directly by its own outputs, having a specified boundary based on inclusion of all significant feedbacks. Human body is a system with a boundary (e.g., the skin) enclosing a number of components (heart, lungs) that interact (the heart pumps blood to the lungs) for a common purpose (to maintain and operate the living body). Collection of unrelated items does not constitute a system. A bag of pebbles is not a system because with addition or removal of a pebble, a bag of pebbles remains and is almost not affected completely by that change. Pebbles behave as a whole if the whole bag is influenced, for example by dropping it; the constituent parts go their own ways with its bursting. It is the property of the system that mainly matters and it may be summarized in the phrase 'behaviour as a whole in response to stimuli to any portion'.

Systems approach

In system approach, there is linkage of all the components and activities affecting each other. It is not wise to have a look at one component by itself without recognizing what it does and what happens to it will affect other parts of the system. Think what happens when you stub your toe, there is reaction in the whole body and different parts may respond differently. Eyes may water, voice may make appropriate sounds, the pulse rate may increase and hands may try to rub the damaged toe. It would be very rash to have alteration in any component of a system without its consequences and reactions elsewhere. You cannot improve a bus/tractor (system) by doing research on one wheel and then making it rather bigger or smaller than the rest or enhance the power and size of the engine without considering the ability of the chassis to support it. These things are common sense in such familiar contexts- these are also applicable to biological and agricultural systems. In agriculture, management practices were usually formulated for individual crop. However, farmers are growing various crops in different seasons depending upon their adaptability to a particular season, domestic requirement and net profit. Therefore, production technology or management practices should be developed and recommended considering all the crops grown in a year or more than one year if any sequence or rotation extends beyond a year. Such a package of management practices for all crops leads to efficient utilization of costly inputs and also reduces production cost. For instance, residual effect of manures and fertilizers applied and nitrogen fixed can considerably bring down the production cost by considering all the crops than individual crops.

Farming system

The process of harnessing solar energy in the form of economic plant and animal product is called farming. A set of inter-related practices/processes organized into a functional entity is called a system. Farming system is a set of agricultural activities organized while preserving land productivity, environmental quality and maintaining desirable level of biological diversity and ecological stability. Farming system is a complex inter-related matrix of soil, plants, animals, implements, power, labour, capital and other inputs controlled in part by farm families and influenced by varying degrees of political, economic, institutional and social forces that operate at many levels. It is a set of elements or components that are interrelated which interact among themselves. Farmer exercises control and choice regarding the type and result of interaction.

It represents integration of farm enterprises such as cropping systems, animal husbandry, fisheries, forestry, sericulture, poultry etc for optimal utilization of resources bringing prosperity to the farmer. The farm products other than the economic products, for which the crops are grown, can be better utilized for productive purposes in the farming systems approach.

Farming systems concept

In farming system, the farm is viewed in a holistic manner. The farmers are subjected to many socio-economic, biophysical, institutional, administrative and technological constraints. A combination of one or more enterprises with cropping when carefully chosen planned and executed, gives greater dividends than a single enterprise, especially for small and marginal farmers.

Farm as a unit is to be considered and planned for effective integration of the enterprises to be combined with crop production activity, such that the end-products and wastes of one enterprise are utilized effectively as inputs in other enterprise. For example the wastes of dairying viz., dung, urine, refuse etc are used in preparation of FYM or compost which serves as an input in cropping system. Likewise the straw obtained from crops (maize, rice, sorghum etc) is used as a fodder for dairy cattle. Further, in sericulture the leaves of mulberry crop as a feeding material for silkworms, grain from maize crop are used as a feed in poultry etc. The selection of enterprises must be based on the cardinal principle of minimizing the competition and maximizing the complementarily between the enterprises.

Sustainability is the objective of the farming system where production process is optimized through efficient utilization of inputs without compromising on the quality of environment with which it interacts on one hand and attempt to meet the national goals on the other. The concept has an undefined time dimension. The magnitude of time dimension depends upon ones objectives, being shorter for economic gains and longer for concerns pertaining to environment, soil productivity and land degradation.

Principles of farming system

- Minimization of risk
- Recycling of wastes and residues
- Integration of two or more enterprises
- Optimum utilization of all resources
- Maximum productivity and profitability
- Ecological balance
- Generation of employment potential
- Increased input use efficiency
- Use of end products from one enterprise as input in other enterprise

Characteristics of farming system

1. Farmer oriented & holistic approach
2. Effective farmer's participation
3. Unique problem solving system
4. Maintains the long-term biological and ecological integrity of natural resources
5. Sustain a desirable level of support to a farm's, communities or regions social, political, and economic well being
6. Dynamic system
7. Gender sensitive
6. Responsible to society
8. Environmental sustainability
9. Location specificity of technology
10. Diversified farming enterprises to avoid risks due to environmental constraints
12. Provides feedback from farmers
13. Enhances quality of life.

Integrated Farming System-Historical background

Integrated Pest Management can be seen as starting point for a holistic approach to agricultural production. Following the excessive use of crop protection chemicals, first steps in IPM were taken in fruit production at the end of the 1950s. The concept was then further developed globally in all major crops. On the basis of results of the system-oriented IPM approach, models for Integrated Crop Management were developed. Initially, animal husbandry was not seen as part of such integrated approaches.

In the years to follow, various national and regional initiatives and projects were formed. These include LEAF (Linking Environment and Farming) in the UK, FNL (Fördergemeinschaft Nachhaltige Landwirtschaft e.V.) in Germany, FARRE (Forum des Agriculteurs Responsables Respectueux de l'Environnement) in France, FILL (Fördergemeinschaft Integrierte Landbewirtschaftung Luxemburg) or OiB (Odling i Balans) in Sweden. However, there are few if any figures on the uptake of Integrated Farming in the major crops throughout Europe for example, leading to a recommendation by the European Economic and Social Committee in February 2014, that the EU should carry out an in-depth analysis of integrated production in Europe in order to obtain insights into the current situation and potential developments. There is evidence, however, that between 60 and 80% of pome, stone and soft fruits were grown, controlled and marketed according to "Integrated Production Guidelines" in 1999 already in Germany for example. In the UK, 22% of fresh fruit and vegetables are grown to Integrated Farm Management standards as recognized by the LEAF Marque.

Animal husbandry and Integrated Crop Management (ICM) often are just two branches of one agricultural enterprise. In modern agriculture, animal husbandry and crop production must be understood as interlinked sectors which cannot be looked at in isolation, as the context of agricultural systems leads to tight interdependencies. Uncoupling animal husbandry from arable production (too high stocking rates) is therefore not considered in accordance with the principles and objectives of Integrated Farming. Accordingly, holistic concepts for Integrated Farming or Integrated Farm Management such as the EISA Integrated Farming Framework, and the concept of sustainable agriculture are increasingly developed, promoted and implemented at the global level.

Related to the 'sustainable intensification' of agriculture, an objective which in part is discussed controversially, efficiency of resource use becomes increasingly important today. Environmental impacts of agricultural production depend on the efficiency achieved when using natural resources and all other means of production. The input per kg of output, the output per kg of input, and the output achieved per hectare of land—a limited resource in the light of world population growth—are decisive figures for evaluating the efficiency and the environmental impact of agricultural systems. Efficiency parameters therefore offer important evidence how efficiency and environmental impacts of agriculture can be judged and where improvements can or must be made.

Against this background, documentation as well certification schemes and farm audits such as LEAF Marque in the UK and 33 other countries throughout the world become more and more important tools to evaluate and further improve agricultural practices. Even though being by far more product or sector-oriented, SAI Platform principles and practices and GlobalGap for example pursue similar approaches.

Following first developments in the 1950s, various approaches to Integrated Pest Management, Integrated Crop Management, Integrated Production and Integrated Farming were developed worldwide (in Germany, Switzerland, US, Australia, and India, for example). As the implementation of the general concept of Integrated Farming and its individual components always should be handled according to the given site and situation instead of following strict rules and recipes, the concept is virtually applicable and being used to various degrees all over the world.

The Farming Systems Prior to the mid-1960's, active research collaboration between technical agricultural scientists (i.e., mainly working on experiment stations), agricultural economists (i.e., mostly in planning units) and anthropologists/ rural sociologists (i.e., generally in academia), was limited. By the mid-1960s, the Green Revolution was beginning to have a major impact on crop production in parts of Asia and Latin America through the introduction of fertilizer-responsive, high-yielding varieties of rice, wheat, and maize in favorable and relatively homogeneous production environments where there was assured soil moisture, good soils, ready access to cheap fertilizer, and relatively efficient output markets. However such conditions did not exist in most of Sub Saharan Africa and in certain parts of Latin America and Asia, and as a result, these areas were bypassed. The reductionist approach failed in terms of developing technologies for resource-poor farmers in less favorable heterogeneous production environments or agricultural areas. This led to the incorporation of a systems perspective in the identification, development, and evaluation of relevant improved technologies. Hence in the mid to late 1970s, the farming systems research (FSR) approach evolved, a basic principle of which was the need to create new types of partnerships between farmers and technical and social scientists. FSR thus became very popular with donor agencies, to the extent that, by the mid 1980s, about 250 medium- and long term externally funded (i.e., in addition to those domestically funded) projects worldwide were implementing FSR-type activities. Between 1978 and 1988, USAID alone had funded 76 bilateral, regional, and centrally funded projects containing a farming systems orientation. Forty-five of these were in Africa. Most of these projects supported the establishment of separate FSR units, which often were poorly integrated into, or poorly linked to, mainstream technology development activities. Although it is probably true to conclude that few of these projects succeeded in producing new technologies that were widely adopted, the approach of looking at farmers' constraints and needs for technical change from within was eventually mainstreamed into most national and international agricultural research programs by the late 1980s. Therefore although donor support for supporting explicit FSR activities dwindled towards the end of the 1980s, most national agricultural research systems (NARS) had adopted major components of the FSR philosophy and approach, and the spirit of the FSR approach lived on. Since then there has been considerable evolution in the methodologies employed (e.g., new farmer participatory research (FPR) techniques, gender analysis, environmental impact analysis, and statistical techniques adapted to on-farm research). Also participation has been broadened to include a wider set of agricultural stakeholders, including extension, development, and sometimes even planning/policy staff. Perhaps even more significant has been incorporating the underlying principles of the farming systems approach into the priorities of donors and nationally based agricultural programs. These include increasing emphasis on participatory approaches and empowerment of farmers and their families and a new focus on ecological sustainability and sustainable livelihoods. Although appropriate technologies still are viewed as important catalysts for improving farmers' welfare, the criteria for relevancy have become more clearly defined and specific. In this discussion it is summarized how the farming systems approach that has evolved over the last 40 years with a very brief indication of the factors that contributed to bringing about those changes. A key dimension of that evolution has been the way the scope or inclusiveness of a systems perspective has been expanded systematically over time. "The way in which the systems perspective is implemented, in particular the scope or inclusiveness of the systems analysis, depends on how, for any given problem, researchers define the ratio of variables to parameters; or put another way, which factors are considered endogenously determined and thus subject to analysis and modification, and which are taken as exogenously determined constants. Because of the analytical difficulties of simultaneously handling large numbers of variables, most of the early FSR programs took only incremental steps away from traditional reductionist approaches by limiting the number of variables they studied and by regarding the other factors that influence the farming system as parameters or constants. As analytical methods have grown more sophisticated, and particularly as farmers have become active partners in the analysis, the ratio of variables to parameters has increased, and the analytical domain has expanded considerably. This evolutionary process is operationally summarized in four phases: 1960's to early 1970's – Farm

Management The farm management approach of the early 1900s was in many ways analogous to practiced, the focus of questions soon shifted to why formally recommended technologies were adopted so rarely. Thus many agricultural economists, particularly those associated with research stations in Africa, Asia, and Latin America, began to evaluate recommended technologies (i.e., usually packages and typically crop-oriented). Prior the mid-1960s, very few station-based experiments were subjected to any economic analysis, and therefore it was not surprising that the conclusion that often emerged was that many existing recommendations were poorly designed or irrelevant, especially when criteria relevant to farmers were applied. In addition three other significant insights emerged:

- Contrary to expectations of many, farmers were found to be natural experimenters, using informal methods and consequently it was wrong to conclude they were conservative and averse to change.
- Farmers' production environments were found to be much more heterogeneous than had been thought, and consequently there was a need to develop technological components that could be adjusted easily and combined variously to better respond to location-specific needs, rather than relying on a few technological packages (i.e., the one size fits all syndrome!).
- And although the recommended technological packages were sometimes compatible with the biophysical environments within which farmers operated, farmers were often not able to adopt them because of their incompatibility with the socioeconomic environment within which they operated. Questions started arising as to whether the current process for developing and evaluating technologies was relevant for resource-poor farmers operating in less favorable and highly variable environments. It also became apparent that standard conventional economic criteria didn't ensure identification of a relevant technology (e.g., farmers and their households had goals other than profit maximization, there were usually multiple market failures for capital, labour, land, and information, and risk and uncertainty were significant issues). As a result many of us came to the conclusion that:
- The neoclassical economic paradigm was ineffective in dealing effectively with all the issues relating to small-scale farmers.

Recent Advances in Integrated Farming Systems

The approach was also too static and deterministic in its orientation rather than recognizing that farmers operate in a dynamic and often uncertain environment.

- The approach of extracting data from farmers (i.e., treating them as objects) and analyzing it independently was much inferior to an approach that recognized the benefits of synergism as a result of interaction and active participation on the part of farmers themselves.
- The approach was flawed in its ex post orientation of focusing on evaluating available technologies, rather than using one that encouraged an ex ante involvement of farmers in the technology design and development process itself. As a result many argued for that the conventional research paradigm needed to be modified drastically and replaced by one that would involve farmers as stakeholders from the beginning of the technology design process and that would use an interdisciplinary approach. Thus momentum developed for the evolution of a new approach based on changing from a "top-down" ("supply-driven") approach to farmers, to one characterized as being "bottom-up" ("demand driven") from farmers.

Late 1970s to early 1980s – Early Farming Systems Approaches given what was discussed in the preceding section it is not surprising that the newly christened FSR activities were focused primarily on technology development objectives. In the anglophone countries, several of the international agricultural research institutes (IARCs) (i.e., especially IRRI and CIMMYT) associated with the Consultative Group on International Agricultural Research (CGIAR) played important roles in the early methodological development and popularization of FSR. At the same time FSR was introduced, with donor support, in many nationally sponsored agricultural research institutes in low-income countries. The FSR approach that evolved was based on the notion that: one had to begin with understanding the problems of farmers from the perspectives of farmers; and that solutions had to be based on a proper understanding of their objectives and their environments, including both biophysical and socioeconomic components. Also a central tenet of the new approach was that not only did farmers have a right to be involved in the technology development and evaluation process, but that their inputs were essential. Other significant features were its holistic perspective, the fact that scientists involved in the process should represent both technical and social scientists, and that the process was by nature iterative. Consequently, some of the Recent Advances in Integrated Farming Systems characteristics of the early farm-management approach started to reappear, leading Johnson to observe, in commenting on the new FSR methods, "there has been much reinventing of the wheel in developmental thinking". Although there was a commitment in principle to include a broader set of farmer-based criteria, in its earliest days FSR continued to focus on how yields of particular crops could be increased. However, even though the new on-farm research approach did involve the inclusion of socioeconomic elements, and hence had a farming systems perspective, it was done generally with a predetermined focus that targeted the productivity of a particular commodity. Thus, this approach involved looking at one part of an enterprise or one specific enterprise and identifying improvements within that focus that were compatible with the whole farming system. For example, CIMMYT and IRRI worked on maize-, wheat and rice-based systems, which were compatible with their crop mandates. Undoubtedly they had a favorable impact in introducing more of a systems perspective to the influential commodity-based research programs that had a strong reductionist orientation. They believed that the predetermined focus approach was directly relevant to farming systems dominated by one crop, because improving the productivity of that enterprise would have the greatest impact on the productivity of the overall farming system. These two IARCs, in particular, played influential roles through networks and training programs in Africa and Asia, thereby exposing scientists to the principles of the farming systems approach. Although the IARCs undoubtedly played an important role in introducing and nurturing the farming systems perspective within national agricultural research systems (NARS), simultaneous independent efforts occurred in developing and promoting the approach, usually supported by donor funds. Notable examples were ICTA (Guatemala), Changmai (Thailand), Unite Experimentales, ISRA (Senegal), and the Institute of Agricultural Research, Ahmadu Bello University, Northern Nigeria. Because of the multi-commodity mandates of most NARSs, the farming systems efforts evolved quickly towards a more holistic orientation (i.e., what is labelled farming systems with a whole farm focus). This approach enabled focusing on constraints in any enterprise depending on farmers' articulated needs. This evolution was further stimulated by two other factors, specifically:

- A desire to encourage greater participation by farmers through addressing their specific needs rather than simply trying to "fit" technologies to specific enterprises that had been preselected by researchers/mandates.
- The trend towards establishing separate area-based farming system teams in contrast to an onfarm testing component associated with each station-based commodity team. The dominant disciplines in the early days of the farming systems approach were cropping systems agronomy and agricultural economics and the methodologies reflected this discipline mix. The trend away from treating farmers as "objects" to treating them as "people", with whom useful interactive dialogue could be established, was helped greatly by increasing reliance on informal surveys or rapid rural appraisal (RRA) techniques. The results of these were sometimes verified by limited-visit formal surveys. The results were helpful in designing on-farm trials that were either managed by researchers or farmers — but usually executed by the latter — and were validated by conventional statistical techniques. Trials

superimposed as separate plots on farmers' operational fields also became commonplace. Three very positive results from these early experiences with the farming systems approach:

- Technical scientists were increasingly sensitized to the complexity and variability of farmers' production environments (i.e., consisting of both physical and socioeconomic components) thereby helping reorient technology generation towards addressing needs of different types of farmers and emphasizing more flexible technological components rather than simply blanket type package technologies.
- The approach provided an opportunity for technical and social scientists to cooperate in the diagnosis of farmers' situations and in the design, testing, and evaluation of new technologies, thus helping those concerned to better understand better the disciplinary perspectives and tools of other specialties.
- Results also demonstrated the importance of complementary policy/support systems (e.g., input distribution systems and product markets) in determining the appropriateness of new technologies. However, a number of limitations or weaknesses became increasingly apparent. Five particularly significant weaknesses were:
- Farmers' participation was still limited largely to roles assigned by researchers and methodologies for obtaining and systematizing farmers' knowledge and for analyzing the results of on-farm experiments were also poorly developed, often resulting in scepticism about how valid were researchers' interpretation of information obtained from farmers.
- Although the complexities of linkages between farm and household in influencing decision making and flows of resources and benefits were increasingly recognized, methodologies for incorporating such considerations into technology design and evaluation were inadequate and too often purely subjective and adhoc in nature.
- Although some linkages had been developed between the different disciplines (i.e., mainly agronomy and agricultural economics) there was still a lot of room for improvement. The application of the farming systems approach to livestock enterprises was generally particularly weak.
- The most commonly used methodologies for data collection and analysis generally were based on the assumption of a monolithic household that could be described by a single objective function and with the household head at the center of the information nexus, in spite of increasing evidence to the contrary (e.g., multiple decision makers, differentiation in terms of distribution of benefits).

- Factors relating to the policy/support system were treated as parameters within which the search for improved technologies took place. This was partly because the mandates of the technology-oriented institutions in which most farming-systems-related work was – and for that matter still is — based did not include objectives of influencing the policy context and support systems. As a result this severely constrained the types of technologies that could be developed/evaluated. Fortunately, there was increasing recognition of the above limitations as the decade of the 1980s progressed. The changes that occurred during the decade were supported by an increasing acceptance of a new developmental paradigm, which characterizes as a "learning process" (i.e. people centered) approach to the earlier "blueprint" (i.e. technology) approach.

Late 1980s and early 1990s – New directions in the farming systems approach

During the mid to late 1980s, some significant methodological and institutional innovations were introduced in the implementation of the farming systems approach with a whole farm focus. Three of the methodological innovations were as follows:

- The development of participatory rural appraisal (PRA) techniques provided a way of responding to three concerns namely: how farmers interpreted their production situations; how this influenced the way they articulated their constraints and needs to researchers; and the desire for farmers to contribute more directly and creatively to the design and evaluation of new technologies. Basically, PRA techniques improved the potential usefulness of farmers' participation not only from the farmers' but also from the researchers' perspective by improving systematization of farmers' knowledge and opinions. Thus researchers could move from working relationships with farmers that were contractual or at best consultative to those that were more consultative and collaborative.
- Techniques for examining intra-household relationships also evolved and as a result increased sensitivity to gender issues. This helped dispel the notion that the farming household was a single decision making unit in which all persons benefited equally from the fruits of technological change. Thanks to the commitment of Feldstein, Poats, Jiggins, Flora and many others methodologies for incorporating gender-related issues began to be mainstreamed into FSR-related activities starting in about the mid 1980s.
- Significant progress also was made in developing more appropriate methods to analyze the results of on-farm research and make recommendations based

on them. Two approaches that deserve special mention are: adaptability (formerly, modified stability) analysis which has proved to be a particularly valuable statistical tool for analyzing results of on-farm trials, particularly those that involve farmer implementation, or both farmer management and implementation; and PRA techniques – especially matrix ranking and scoring — which have enabled farmers' evaluative criteria to be systematically taken into account not only in designing on-farm trials but also in evaluating their results. As a result of these methodological innovations, the potential usefulness of farmer participation in all phases of the research process, from problem identification to technology design, development, and evaluation, has been improved. However, there have also often been institutional changes to encourage more collaborative and collegial relationships between farmers and researchers. Four examples of how new institutional arrangements have been used to promote such relationships are the following:

- Farmer groups (both formal and informal) have increasingly been used to help empower farmers and improve the efficiency of the research/development process by providing a focal point (hence potentially improving the multiplier effect) for interaction with farmers and by facilitating farmer-to-farmer interaction. Given the right conditions, such groups have been proven to be very useful in influencing the research agenda and in evaluating technologies (e.g., in Botswana, Tanzania, Mali, Zambia).
- Another less widely used, but potentially even more powerful, way of empowering farmers — thereby facilitating the development of collaborative/collegial working relationships — is to involve farmers groups in the decision to allocate research funds, thereby ensuring that the agricultural research agenda is focused more tightly on farmers' real needs. Ashby of CIAT has been one of the pioneers of this approach through the formation of CIALS (Comites de Investigacion Agropecuaria Local) in Colombia. Somewhat analogous approaches are also being tried in some West African countries (e.g., Mali, Senegal).
- Until relatively recently, the conventional wisdom was that farmers' participation in the technology development process should be confined to the adaptive end of the research spectrum but thanks to the pioneering efforts of Sperling and colleagues working with beans in Rwanda, farmers have been found to be able to make uniquely valuable contributions toward the development of improved varieties. As a result, participatory plant breeding activities have become increasingly popular in a growing number of IARCs (e.g., beans in CIAT, maize in CIMMYT, pearl millet in ICRISAT, barley in ICARDA, and cassava in CIMMYT) as well as in a few NARSs. The number of agricultural development actors has expanded in recent years

from farmers and public sector agencies to include private sector profit-oriented entities as well as NGOs. The importance of interactive linkages between the developmental stakeholders if relevant technologies were to be designed, disseminated, and adopted, was recognized in the early days of FSR. Thus the establishment of committees at the regional/district level consisting of representatives of the different stakeholders, often including farmer representation, for information exchange, enhanced coordination, and design of collaborative initiatives has become more commonplace. Decentralization policies in terms of governance and in some countries local approval of technological recommendations (i.e., thus helping ensure that new technologies are more likely to address the needs of specific farmers) have also helped this process. Consequently, major improvements have occurred in the application of farming systems with a whole-farm focus from the mid-1980s until the present time. Also, the range of social (e.g., anthropological, sociological) and technical (e.g., pest-, disease-, livestock-related) disciplines associated with the application of the farming systems approach has broadened, and farmers and their households have increasingly played more influential roles in all aspects of the technological research process. Other major changes have involved developing technological options rather than standardized packages for farmers and increased transparency in providing information on the conditions in which technologies are most likely to fit and perform best (i.e. targeting information) and suggested fallback strategies if they are not applied in the optimal manner (i.e. conditional information). Most current FSR efforts of NARS fall within the whole-farm focus phase of the farming systems evolutionary ladder, although the degree to which the methodological and institutional issues discussed in this section are applied varies greatly. However, another major limitation that became increasingly apparent towards the end of the 1980s related to concerns about ecological sustainability and environmental degradation. Farmers often appreciate that some of their agricultural practices may contribute to environmental degradation, but short-term survival considerations can lead them to pursue strategies that ensure short-run food supplies but degrade the environment and reduce longer run production potential (e.g., resource-poor households being forced to cultivate marginal soils to meet their subsistence needs or to intensify cropping systems without the means to purchase the inputs necessary for soil fertility maintenance). However, evidence is also increasing that higher income farmers practicing high-input cropping also cause significant environmental damage (e.g., decreasing productivity in the intensive rice-wheat systems of South Asia). Although researchers often perceive or foresee ecological degradation to be a problem, farmers for reasons just given, may not mention such concerns, unless they threaten immediate survival. Thus, such concerns may not be addressed adequately in FSR, as it was originally conceptualized. Because felt needs articulated by farmers are likely to be strongly biased towards short-run productivity, researchers and development practitioners are increasingly concerned about possible conflict between strategies designed to improve short-run productivity and those aimed at ensuring long-run ecological sustainability. As a response to this, the principles of the farming systems approach and RRA/PRA have increasingly addressed ecological sustainability — hence the term farming systems with a natural resource systems focus. Incorporating the farming systems approach into natural resource-related issues has taken two major directions:

- One has involved development and implementation of methodologies to work with farmers to assess bioresource (nutrient) trends and flows at the farm-household level. The idea is to determine biomass trends over time and to help identify, with the farmer, vulnerable parts of the farming system for which modified strategies are needed to promote ecological sustainability. Many of these techniques have their origin in RRA/PRA. Lightfoot and others at ICLARM developed the principles for the approach, which involved integrating aquaculture and agriculture. A somewhat analogous approach has been developed in Africa, also involving extensive interaction with farmers to derive bio resource flows on their farms and to identify strategies for reversing any trends towards ecological degradation. These approaches help promote ecologies problems from a foreseen problem to a felt problem in the eyes of the farmers and depend heavily on farmer empowerment and on interactive collaborative/collegial relationships with researchers and development practitioners. The solutions often are farm-specific and involve changes in practices, and, as such, implementation is primarily the responsibility of farmers and their households. Hence, farmers need to buy into adjustments via the participatory identification and design stages.
- The other is eco regional research, which has been undertaken by some IARCs in collaboration with national research and development institutions. The goal has been to mobilize and focus CGIAR and national funding on natural resource management research with the twin objectives of improved productivity and environmental conservation. Teams of international and national scientists conduct collaborative research targeted at priority regional problems, the aim being to achieve critical mass and economies of scale in the conduct of strategic and applied research, closely linked to adaptive research to address location-specific conditions. The approach involves, in a collaborative operational mode, all the stakeholders involved in the agricultural development process. There are at least three major challenges when

implementing farming systems with a natural resource focus. These are: that large investments are required to address complex processes that are manifested differently across locations; a long time frame needed to improve ecological sustainability and assess progress; and finally, because of the precarious existence of many farmers and their households, ecological sustainability initiatives are likely to be attractive only if they simultaneously improve short-run welfare. Currently – A Sustainable livelihood focus during the 1970s and 1980s, the scope of FSR was broadened systematically to encompass a wider set of issues. That is, the number of variables included within the analytical and prescriptive domain was increased with each new development. The sustainable livelihoods (SL) can be viewed as the end product (i.e., final phase or phase four) of that process. Although many SL components are similar to earlier farming systems approaches, there are several distinct features. For example:

- Livelihoods are not just jobs but rather comprise a situational determined complex of activities, asset sets, entitlements, and social relationships that are managed by households to assure their minimum basic needs over time. Because these vary across households in different socio-economic strata within the community, the livelihood strategies of different households also vary. The SL approach usually places greatest emphasis on the most vulnerable households that are in or near poverty and experiencing either chronic or temporary food insecurity.
- The SL approach places equal emphasis on enhancing efficiency through improved productivity, achieving greater social equity through poverty alleviation, and protecting and enhancing resource base productivity. Thus the SL approach requires a combination of analytical methods including conventional FSR, political economics, anthropology, and environmental science. Generally, these are applied through interdisciplinary teams working in close partnership with local communities.
- The SL approach explicitly links technical change at the household level with complementary changes at the meso- and macro-levels. Time and space does not permit a detailed discussion of the SL methodology but the aim is to empower households and communities by strengthening their abilities to combine indigenous and modern knowledge, analyze situations, define problems, identify opportunities, conduct experiments, and formulate plans of action. A distinctive feature of the SL approach is the emphasis placed on designing interventions that simultaneously improve current and future productivity, reduce poverty, and protect the environment, without weakening and preferably strengthening the coping and adaptive strategies of the most

vulnerable groups in the community. Successful implementation requires major changes in the roles traditionally played by researchers and farmers. It has been suggests that outside professionals can help catalyze farmers' empowerment by acting as conveners of farmers' meetings; facilitating exchange of experiences among farmers; and supplying farmers with new technical options and new methodologies to elicit, systematize, and utilize traditional knowledge to solve new problems. Obviously new production technologies should improve short-term productivity, fit the objectives and resource constraints of poor households, and protect the long-term productivity of the resource base. However, these are not new ideas but what the SL approach does add is the notion that improved technologies should be consistent with and, if possible, reinforce the coping and adaptive mechanisms of resource-poor farmers. Using this idea at least three properties of new technologies that would satisfy this are:

- The notion of flexibility meaning that new technologies should, wherever possible, increase farmers' flexibility to adapt their production and broader livelihood systems to stochastic shocks and to the constantly changing economic environment.
- Emphasis needing to be placed on technologies that reduce risk (e.g., new more resistant/tolerant crop varieties and agronomic practices that reduce the impact of biotic and abiotic stresses, technologies that promote enterprise diversification).
- Finally, new technologies should complement, and not conflict with, the complex livelihood systems of poor households (e.g., if technologies require changes in the demand for labor, such changes may increase returns to labor for the poor in own-farm, wage labor, or off-farm activities, may be neutral, or may compete with labor requirements in their current set of enterprises). Although it is obviously elegant and holistic, the application of SL to date is limited. There are a number of important challenges for wide application. A particular problem is the problem of upscaling: because livelihood systems and their associated coping/adaptive strategies vary greatly across sites, and thus the results of SL field work usually are very location specific; and the full application of the SL concept requires well trained and experienced interdisciplinary teams that are skilled in participatory methods and that can work together over extended periods. Cropping systems at present mainly in the rice-wheat cropping system biological research is focusing on issues related to natural resource management (NRM). Its most notable success to-date has been the recent development of several resource conservation technologies (RCTs) due to the efforts championed by rice and wheat coordinating (RWC)

units with its NARS (National Agricultural Research System) partners, including the private-sector machinery manufacturers.

Resource Conservation Equipment & Technology

Equipment	Technology
Laser land leveler	30-50% saving in water
Rotavator	50% fuel saving & better quality seed bed
Zero till drill/minimum till drill/ multipurpose tool bar/ raised bed planter	5-10% increase in yield and saving of Rs. 2000-3000/ha
Pressurized irrigation	20-30% saving in water
Rotary power weeder	20-30% saving in time and labour
Vertical conveyor reaper/ combine	Timely harvesting, more yield
Multi-crop thresher	50% saving in labour and time and 54% saving in cost of threshing
Straw combine	Recovers 50% straw and also 70-100 kg grain/ha resulting into an average saving of Rs. 1250/ha.
Straw baler	Makes bales and checks environmental pollution
Straw cutter-cum-spreader	Cuts and spreads the straw evenly and helps in sowing by zero till drill
Improved manual harvester for mango & kinnow	No damage to fruit and higher capacity

The focus on RCTs is important for reasons other than efficiency and sustainability per se. The new RCTs provide a novel 'platform' for land and water management approaches and to introduce new crops and varieties into the systems, which may also help to re-establish better ecological balance. The biophysical and socio-economic heterogeneity in different IGP transects must be borne in mind in planning future programs. In the west, traditionally a wheat-based production system, introduction of intensive rice cultivation has raised concern about environmental sustainability due to antagonism between the current soil-water production requirements of the two crops. The challenge for RWC is to undertake research to determine what possibilities exist to grow rice in different ways to the benefit of the RWSs in terms of productivity, diversity and sustainability (particularly of water use) and determine under what circumstances (including national policies) such changes are appropriate. The RWC can make significant contributions both by improving water use efficiencies at farm-level through new RCTs, including laser land leveling and bed planting, and by joining with the CGIAR's Challenge Program on Water and Food. In the east, where the production systems are traditionally rice-based, intensification and diversification in the winter (non-monsoon) season will need to be focused on enhancing economic viability, learning from farm-level experiences with diversification in Bangladesh. The needs for expansion of successful RCTs, for system diversification and for water management research present an attractive window of opportunity for adoption of such a strategy and for exploring different options for securing medium-term funding. At the same time, the RWC members should also examine a move towards a more equitable cost sharing arrangement in line with their size, degree of involvement and capacity to bridge the gap in sustainable funding for the CU. Conservation agricultural (CA) practices are widely adopted for rainfed and irrigated systems in tropics/subtropics and temperate regions of the world. Acreage of CA is increasing steadily worldwide to cover about 108 m ha globally (7% of the world arable land area). Recent estimates revealed that CS based RCTs are being practiced over nearly 3.9 m ha of South Asia. Site specific crop management practices including SSNM, BMP besides IPNS results in better crop yield and efficiency. Building crop management programmes by integrating available BMPs for the site with due consideration of interactions and their careful management result in huge cumulative benefits – additive, and/or multiplicative from the biological system of crop production. This would greatly reduce external input supply and achieve highest fertilizer use efficiency and sustainability. With increased productivity, acreage under arable cropping can be reduced, fertilizer requirement rate can be reduced, residual accumulation and further wastage of nutrients can be reduced, more C can be fixed due to the higher biomass and higher oxygen to atmosphere can be released – a true sustainable meaning.

In recent years, the adoption of various precision agriculture technologies by U.S. maize and soybean growers has been increasing dramatically. Among the more commonly adopted precision farming technologies are those associated with auto navigation of equipment, autocontrol of individual rows or nozzles of application equipment, and variable rate control of application equipment. Interest in data management technologies is also growing as these capabilities become more available and affordable. The increasing availability and affordability of sub-inch (cm) accuracy in GPS receiver technologies has contributed to growers' interest in auto navigation and auto control rather they offer the potential to reduce production costs through more accurate placement of crop inputs. An intangible, yet meaningful, benefit of automat technologies is the reduction in operator fatigue accompanied with the freedom to monitor equipment more closely. The opportunity for variable rate (VR) control for application of certain crop inputs, primarily fertilizers and seeds, has been available for some time to U.S. agricultural retailers and increasingly so at the farmer level in recent years. Fertilizer retailers in the U.S. adopted VR application technologies for application of phosphorus and potassium fertilizers plus lime as long ago as the 1990's. Variable rate application of phosphorus and potassium fertilizers plus lime has proven effective in reducing whole field input costs and improving yields in areas of fields with serious deficiencies for P, K, or soil pH. One of the reasons this VR technology has been effective is that the agronomic data to support the

VR decisions can easily be generated from spatially intensive soil sampling data that are then used to develop spatial application maps. In other words, the VR decisions for these inputs are based on a sound understanding of the agronomic relationships between a single variable (spatial soil sample data) and crop response. While effective, these particular VR technologies do not result in "quantum leap" shifts in the annual rate of maize productivity. At best, VR application of P, K, and lime improve the lower yielding areas of fields identified as deficient for these important nutritional factors. Variable rate control technology is also available for nitrogen fertilizer applications or seeding rates. Both technologies are becoming increasingly available to farmers and becoming common standard accessories on applicators and planters. Active optical reflectance sensors are very sensitive to changes in plant biomass and overall health of a maize crop. Various reflectance indices (e.g., NDVI, chlorophyll index) have been shown to closely correlate with N content of plants or overall N uptake by crops and also correlate closely with eventual grain yield.

The System of Rice Intensification, or SRI for short, is a fascinating case of rural innovation that has been developed outside the formal rice research establishment both in India and the rest of the world. The System of Rice Intensification (SRI) is new technique of rice cultivation which changes the management of soil, water and nutrients that support optimal growing environment for rice. The SRI showed that keeping paddy soils moist gives better results, both agronomically and economically, than flooding the soil throughout its crop cycle. This benefit is enhanced by complementary agronomic practices that greatly increase the growth of roots and of soil biota which make it possible to grow more productive phenotypes from any rice genotype. The SRI, thus is currently attracting the greatest attention to address the issue of 'more yield with less water'. The SRI can also reduce GHG emission and improves soil health. A study at IARI, New Delhi showed that the global warming potential (GWP) in the SRI was only 28.9% over the conventional method. It increased the water productivity by 44% compared to conventional planting method. The seed rate in SRI was much lower which reduced the input cost. The SRI, therefore, seems to be a 'win-win' technology. One of the most promising developments in the controversial field of genetic engineering is the success obtained in breeding a nutritionally enriched rice variety now popularly being referred to as 'golden rice'. This golden rice is genetically modified rice which contains genes that produce high levels of beta carotene and related compounds. Beta carotene is contained in yellow fruits like carrots (from which it gets its name) and mangoes and in vegetables like spinach. Beta carotene and other related compounds are converted in the human body to the crucially needed vitamin A. Unfortunately, many in the developing world that do not have access to fruits and vegetables, suffer from chronic vitamin A deficiency which results in night blindness. Night blindness plagues millions of undernourished people in Asia, including India, crippling their lives. According to the WHO, vitamin A deficiency hits the poor in 96 countries of the world, resulting in over five lakh blind children every year. This blindness is irreversible, these children will never see. Maize is a major cereal crop for both human and livestock nutrition, worldwide. Protein from cereals including normal maize, have poor nutritional value because of reduced content of essential amino-acids such as lysine and tryptophan leading to harmful consequences such as growth retardation, protein energy mal-nutrition, anemia, pellagra, free radical damage etc. As a consequence, the use of maize as food is decreasing day by day among health conscious people. The complex nature of these problems posed a formidable challenge. This challenge was gladly accepted by two distinguished scientists of CIMMYT, Mexico, Dr. S. K. Vasal and Dr. Evangelina Villegas whose painstaking efforts for a period of 3 decades led to development of Quality Protein Maize (QPM) with hard kernel, good taste and other consumer favouring characteristics. This work is globally recognized as a step towards nutritional security for the poor. QPM research and development efforts appropriately spread from Mexico to Central and South America, Africa, Europe and Asia. India also benefited with such germplasm and developed its first QPM composite variety 'Shakti–1' released in 1997 for commercial cultivation across the country. In 1998, the Rajendra Agricultural University (RAU) stepped ahead with further R & D towards the development of QPM variety and popularization of QPM as food. As of today, RAU has marched ahead and has its own success story upon QPM. It is an improved variety of maize which contains higher amount of lysine and tryptophan with lower amount of leucine and isoleucine in the endosperm than those contained in normal maize. Such balanced combination of amino acids in the endosperm results into its higher biological value ensuring more availability of protein to human and animal than normal maize or even all cereals and pulses. Vertical farming is a concept that argues that it is economically and environmentally viable to cultivate plant or animal life within skyscrapers, or on vertically inclined surfaces. The idea of a vertical farm has existed at least since the early 1950s. Vertical farming discounts the value of natural landscape in exchange for the idea of "skyscraper as spaceship". Plant and animal life are mass produced within hermetically sealed, artificial environments that have little to do with the outside world. In this sense, they could be built anywhere regardless of the context. This is not advantageous to energy consumption as the internal environment must be maintained to sustain life within the skyscraper. The concept of "The Vertical Farm" emerged in 1999 at Columbia University. It promotes the mass cultivation of plant and animal life for commercial purposes in skyscrapers. Using advanced greenhouse technology such as hydroponics and aeroponics, the skyscrapers could theoretically produce fish, poultry, fruit and vegetables. In the recent past greenhouse/polyhouse technology has been successfully popularized and the high value cash crops are being grown by the farmers. Research was undergoing on design and development of polyhouses and development of agro-techniques including post harvest and value addition. The research efforts over last two decades (1989-90 to 2010) under the aegis of AICRP– Cropping System (now AICRP in integrated farming system) have lead to several significant findings (Personal Communications).

The summary of major achievements are listed here under

- Resource efficient alternative cropping systems to rice-wheat system were developed for 11 states viz. Haryana, Punjab, U.P., Bihar, Uttarakhand, parts of Orissa, M.P., W.B., Jammu Kashmir, H.P. and Gujarat, involving potato, maize, onion, sunflower, green gram, brinjal, berseem (F), cabbage, okra, radish, potato etc, which gave average productivity of 20 t/ha as rice equivalent yield as compared to 11-13 t/ha, which led to induce the scope of increasing per unit productivity substantially.
- Likewise, efficient alternatives to pearl millet-wheat system in five states, viz.; Haryana, Gujarat, U.P., Rajasthan and Maharashtra, involving potato, green gram, cowpea, mustard, cluster bean, chickpea etc. were identified, which gave the productivity up to 12 tonnes per ha as compared to existing 8 t/ha.
- Efficient alternatives to soybean-wheat system in three states, viz.; M.P., Rajasthan and Maharastra, were developed. Diversification options involving high value crops like isabgol, groundnut, soybean, potato, turmeric etc resulted in average productivity of 15 tonnes/ha as compared to existing 12 t/ha (rice equivalent) yield.
- Integrated plant nutrient supply systems have been established to be beneficial and standardized. Substitution of 25-50% N with FYM or green manure in rice-wheat system was found to increase the productivity by 4%, which may save chemical fertilizers worth Rs. 7.0 crores. It also helped in increase of soil organic carbon by 55.9%. In rice-rice system, green manuring increased the yield by 3.6%.
- Site Specific Nutrient Management Technology has been proven to be a potential tool in breaking the yield barrier through efficient and balanced nutrient management leading to significant productivity increase. Results clearly show that by adoption of SSNM, across the locations, grain yields of more than 13 t/ha in rice - wheat system (with a contribution of 58% rice and 42% wheat) and 12-15 t/ha in rice-rice system (with a contribution of 48% kharif rice and 52% rabi rice), are achievable. It also helped in increase of organic carbon by 55.9%.

- The use of different resource conservation technologies in rice-wheat system has been established to help in significant improvement of crop productivity and resource saving. Adoption of RCTs lead to an improvement in productivity of rice by 1 to 7% in mechanical transplanting, 3 to 8% in drum seeding, 1 to 2% in zero till drilling and 10 to 13% in system of rice intensification (SRI) compared to traditional hand transplanting at different locations. Similarly, in wheat, productivity increased by 5 to 29% in zerotill drilling, 4 to 11% in strip-till drilling and 3 to 23% in bed planting compared to conventional sowing.
- With the application of recommended technologies and balanced fertilization to at least 10% of the area under different cropping systems, spread over various agroclimatic regions as indicated by on-farm research on cultivator's fields, India can easily add about 5 million tonnes of food grain equivalent to rice to its food bowl.
- Through long-term studies in all the major cereal-cereal cropping systems, it has been proved that continuous use of chemical fertilizers (NPK) only and omitting P or K application leads to occurrence of deficiency of these nutrients and decline in yield of crops. However, the extent of yield reduction and occurrence of P or K deficiency is governed by the inherent soil supplying capacity and cropping system adopted.
- On-farm crop response to N, P and K application in different cropping systems has been determined. The average response of different cropping systems was 8-12 kg rice equivalent yield/kg of any of the major plant nutrients like N, P or K with mean economic responses of 7-10 Rs./Re invested on N, 4 Rs./Re spent on P and 6-8 Rs./Re spent on K.
- To improve the sustainability of rice-wheat system over the years, growing of legumes as break crop showed a marked influence on weed flora over the years and a reduction of 45 to 61% weed count was noted in succeeding rice-wheat crop cycle. To improve upon soil organic carbon and micro-nutrients application of organic manures has been proven very effective.
- In hybrid as well as inbred rice, N management through LCC proved superior to locally recommended N application in three splits and it was found possible to curtail 20-30 kg of fertilizer N/ha without sacrificing rice yield, when N is applied as per LCC values. N application at LCC.

Integrated Farming System - A Review

The growth rate of agriculture in the recent past is very slow inspite of the rapid economic growth in India. According to the Economic Survey of India, 2008, the growth rate of food grain production decelerated to 1.2% during 1990-2007, lower than the population growth of 1.9%. It is projected that in our country population will touch 1370 million by 2030 and to 1600 million by 2050. To meet the demand, we have to produce 289 and 349 mt of food grains during the respective periods. The current scenario in the country indicates that area under cultivation may further dwindle and more than 20% of current cultivable area will be converted for non-agricultural purposes by 2030 (Gill et. al., 2005).

The operational farm holding in India is declining and over 85 million out of 105 million are below the size of 1 ha. Due to ever increasing population and decline in per capita availability of land in the country, practically there is no scope for horizontal expansion of land for agriculture. Only vertical expansion is possible by integrating farming components requiring lesser space and time and ensuring reasonable returns to farm families. The Integrated Farming Systems (IFS) therefore assumes greater importance for sound management of farm resources to enhance the farm productivity and reduce the environmental degradation, improve the quality of life of resource poor farmers and maintain sustainability. In order to sustain a positive growth rate in agriculture, a holistic approach is the need of the hour. Farming system is a mix of farm enterprises in which farm families allocate resources for efficient utilization of the existing enterprises for enhancing productivity and profitability of the farm. These farm enterprises are crop, livestock, aquaculture, agro-forestry, agri-horticulture and sericulture (Varughese and Methew, 2009).

In such diversified farming, though crop and other enterprises coexist, the thrust is mainly to minimize the risk, while in IFS a judicious mix of one or more enterprises along with cropping there exist a complimentary effect through effective recycling of wastes and crop residues which encompasses additional source of income to farmer. IFS activity is focused around a few selected interdependent, inter-related and interlinking production system based on crops, animals and related subsidiary professions.

Integrated farming system approach is not only a reliable way of obtaining fairly high productivity with considerable scope for resource recycling, but also concept of ecological soundness leading to sustainable agriculture. With increasing energy crisis due to shrinking of non-renewable fossil-fuel based sources, the fertilizer nutrient cost has increased steeply and with gradual withdrawal of fertilizer subsidy. It is expected to have further hike in the cost of fertilizers. This will leave the farmers with no option but to fully explore the potential alternate sources of plant nutrients atleast for the partial substitution of the fertilizer nutrients for individual crops and in the cropping systems.

Possible output of integrated farming system

Since Integrated Farming System (IFS) is an interrelated complex matrix of soil, water, plant, animal and environment and their interaction with each other enable the system more viable and profitable over the arable farming system. It leads to produce the quality food. To strengthen the food chain, it is essential to eliminate nutritional disorder which has been realized on account of appearing deficiency of mineral nutrients and vitamins in food being consumed. Horticultural and vegetable crops can provide 2-3 times more energy production than cereal crops on the same piece of land and will ensure the nutritional security on their inclusion in the existing system. Similarly inclusion of bee-keeping, fisheries, sericulture, mushroom cultivation on account of space conservative also give additional high energy food without affecting production of food grains. The integration of these enterprises will certainly help the production, consumption and decomposition in a realistic manner in an ecosystem.

Likewise, it is pre-requisite in farming system to ensure the efficient recycling of resources particularly crop residues, because 80-90% of the micronutrients remain in the biomass. In the Indo-Gangetic plains, where rice straw is not recycled in an effective way and even in Punjab where rice cultivation is practiced on 2.6 m ha produces about 16 m tonnes of paddy straw which is destroyed by burning. To curtail such precious input loss, the use of second generation machinery for efficient crop residue management to conserve moisture, improve soil micro-organism activities, regulate soil temperature, check soil erosion, suppress weed growth and on decomposition improves soil fertility. Its beneficial effect can also be accrued by incorporating with the soil. The crop residue can be used as floor thatch for cattle shed, composting, growing mushroom and for dry fodder also. Multiple use of water for raising crops, fruits, vegetables, and fishery may also enhance the water productivity. Likewise, in villages, the sewerage water can be purified through Hydrilla biomass before its release to fish pond. Besides, the community land in the villages, which are accessible to better use, must be used for productive purpose. Therefore, adoption of concept like social forestry, water harvesting and recycling fishery, and stall feeding to the animals (goatery / piggery) will add to the profit margin with other numerous indirect benefits of employment and improved ecology of the area. Such types of enterprise integration generate additional income varying from Rs 20, 000- 25,000/ha under irrigated and Rs 8,000-12,000 under rainfed ecosystem. The income enhancement due to integration of processing and on-farm value addition by 25-50%, yield improvement on account of improved soil health by 0.5-1.0 tonne/ha, cost reduction by Rs.500 - 1,000/ha and employment generation by 50-75 man days/ household have also been observed (Gill et.al., 2009).

Present status of farming system research

The preliminary investigations clearly elucidated that integration of agricultural enterprises viz., crop, livestock, fishery, forestry etc. have great potential towards improvement in the agricultural economy. These enterprises not only supplement the income of the farmer by increasing the per unit productivity but also ensure the rational use of the resources and further create employment avenues. The following of suitable crop choice criteria having deep and shallow root system, inclusion of legume crop as catch, cover and fodder crops and adoption of bio-intensive complimentary cropping system along with other enterprise will certainly prove as a self sustained production system with least cost of production. The farming system is governed by various forces viz., physical environment, socio economic conditions, political forces under various institutional and operational constraints and above all government favorable policies, which may keep the food security intact and livelihood fully protected.

In traditional Chinese system, the animal houses were constructed over a pond so that animal waste fell directly into the water fueling the pond ecosystem, which the fish could then feast on for food. Not only were the fish harvested but the pond water, now with extra nutrients was used for irrigation in crops. The maximum return (Rs 79,064/ha) was earned from fisheries + piggery + poultry as compared to Rs 5,33,221 from the rice-wheat system and registered 48.6% gain. This also generated additional employment of about 500 man days/ha/ annum (Gill et. al.,2005).

For poor people, it starts small with ducks and chickens; then a few goats are kept for milk or fattening and to slaughter for a day of sacrifice; next a milch cow; then a bullock for ploughing in cooperation with another one buffalo family; then two bullocks. These can be used to plough the fields of others- a very lucrative business in the planting season. In India, one would add a milch buffalo at the apex of desirable animals on the farm. In the Vietnamese concept, the pigs will be the second step in the ladder. The concept means to start with small livestock and women and then the household will step by step get out of poverty. The poorest households kept only poultry and these households were those most dependent on common property resources for their living (e.g. use and sale of firewood from the forest). A similar stratification has been reported in several studies from Asia (Lasson and Dolburg,1995) Survey on farming systems in the country as a whole revealed that milch animals; cows and buffaloes irrespective of breed and productivity is the first choice of the farmers as an integral part of their farming system. However, from economic point of view, vegetables and fruits (mango and banana in many parts of the country) followed by bee keeping, sericulture, mushroom and fish cultivation was the most enterprising components of any of the farming systems prevalent in the country. The average yield gaps between 27 pre-dominant and 37 diversified farming systems were examined across the agro-climatic zones through detailed survey on characterization of on-farm farming systems. Diversification of farming system by integration of enterprises in varied farming situations of the country enabled to enhance total production in terms of rice equivalent yield ranging from 9.2% in eastern Himalayan region to as high as 366% in Western-plain and Ghat region when compared to prevailing farming systems of the region. A number of success stories on IFS models including Sukhomajari Watershed of Chandigarh, Fakot Watershed in hilly areas of Uttarakhand. Jayanthi models for almost all the situations ofTamil Nadu, WTCER model for coastal and irrigated alluvial lands of Orissa, Darshan Singh Model for irrigated conditions of Punjab, PDCSR model, for western Uttar Pradesh. and many more in different parts of the country suggest that farmers' income can be increased manifold by way of diversification of enterprises in a farming system mode for sustainability and economic viability of small and marginal category of farmers.

Productivity enhancement by IFS

In view of serious limitations of horizontal expansion of land for agriculture, only alternative left is vertical expansion through various farm enterprises requiring less space and time but give high productivity and ensuring periodic income especially for the small and marginal farmers. The highlights about the research investigations carried out in India towards farming system outcome are discussed to conceptualize its significance towards farming community livelihood. In a study conducted at ICAR Research Complex, Goa, it was revealed that rice-brinjal crop rotation is the best in terms of productivity and profitability owing to higher yield of the brinjal. The system yielded a total productivity of 11.22 t/ha rice grain equivalent yield with a net return of Rs.46, 440/ha. Further, with the integration of mushroom and poultry production (based on the resources availability within the system) the system productivity was increased to 21, 487 kg/ha especially with rice-brinjal rotation leading to an additional returns of Rs 30,865/ha with integration. In addition, the system approach was found to sustainable as reflected from the changes in soil organic carbon and indicated by sustainability yield index (Korikanthimath and Manjunath, 2009).

In Tamil Nadu, the IFS increased the net return on an average of Rs 31,807/ha/ year over the arable farming (Rs 19,505/ha/year). While in Goa, when coconut was integrated with crop, vegetables, mushroom, poultry and dairy enabled to enhance Rs 17,518/ha/annum over the cashew nut cultivation alone. In Madhya Pradesh, the integrated farming gave a margin in net return of Rs 17,198/ ha/ year over the arable farming. In Uttar Pradesh, the average enhancement in return was Rs 45,736/ha/annum over the existing crop-based farming system.

In Haryana, Singh et al.(1993) conducted studies of various farming systems on 1 ha of irrigated and 1.5 ha of unirrigated land and found that under irrigated conditions of mixed farming with crossbred cows yielded the highest net profit

(Rs 20,581/-) followed by mixed farming with buffaloes (Rs 6,218/-) and lowest in arable farming (Rs 4,615/-). In another study conducted with 240 farmers of Rohtak (wheat-sugarcane), Hisar (wheat-cotton) and Bhiwani (gram-bajra) districts in Haryana which represented zones of different crop rotations revealed that maximum returns (Rs/ha) of 12,593, 6,746 and 2,317 were obtained from 1 ha with buffaloes in Rohtak, Hisar and Bhiwani, respectively. The highest net returns from Rohtak was attributed to the existence of a better soil fertility type and of irrigation facilities coupled with better control measures compared to other zones. In terms of total man days, Rohtak had the highest employment potential followed by Hisar and Bhiwani. The employment potential under mixed farming conditions was predominantly from livestock rather than crop production (Singh et al.,1999).

Another study involving cropping, poultry, pigeon, goat and fishery was conducted under wetland conditions of Tamil Nadu conducted by Jayanthi et al.(2001). Three years results revealed that integration of crop with fish (400 reared in 3 ponds of 0.04 ha each), poultry (20 babkok layer bird), pigeon (40 pairs), and goat (Tellichery breed of 20 female and 1 male in 0.03 ha deep litter system) resulted in higher productivity, higher economic return of Rs 1,31,118 (mean of 3 year). Integration of enterprises created the employment opportunities where in comparison to 369 mandays/year generated in cropping alone system, cropping with fish and goat created additional 207 mandays/annum. The resources were recycled in such a way that fish were fed with poultry, pigeon and goat dropping. Similarly, extra poultry, pigeon and goat manure and composted crop residue of banana and sugarcane were applied to the crops. The four conventional cropping system tried were rice-rice-blackgram, maize-rice-blackgram, maize-ricesunhemp and rice-rice-sunhemp.

Balusamy et al. (2003) explained that rice + Azolla-cum-fish culture is one of the economical option in such type of area. Monoculture system relies mainly on external inputs while in integrated system, recycling of nutrients takes place that help in reducing the cost of production for economic yield. The fish in rice field utilized the untapped aquatic productivity of rice ecosystem as the rice bottom is highly fertilized on account of the production of zoo and phytoplankton and these resources are fully utilized by the fish. The data clearly advocated the beneficial effect of Azolla on rice+fish. The gross income obtained in rice + Azolla + fish was 25.7 % more over the rice crop and 6.9 % more over the rice + fish. The net income followed the same trend. Thus rice + Azolla + fish on an average gave Rs 8,817/ha more over the rice monoculture and Rs.3, 219/ha over the rice + fish. This model was proposed for extensive scale adoption in Tamil Nadu.

Farming system is a resource management strategy to avail maximum efficiency of a particular system. Studies conducted at ICAR Research Complex for Goa revealed the higher energy use efficiency of IFS with rice (Manjunath, 2002). The mean total energy input varied considerably among the systems. Integration of poultry and mushroom enterprise with rice-brinjal system required highest energy input (52,030 MJ/ha) and followed by rice groundnut system integrated with mushroom and poultry (46,077 MJ/ha). However, rice cropping alone without any rice based crops or enterprises recorded the least requirement of energy. The energy output was maximum (1,65,334 MJ/ha) under rice-brinjal + mush room + poultry with 3.18 system energy efficiency mainly due to the lesser energy input involved in contrast to energy rich output enterprises. The output of all multi – rice based enterprise was reasonably good varying from 1, 00,911 to 1, 05,627 MJ/ha excluding brinjal crop based farming system. It is thus evident that efficient utilization of scarce and costly resource is the need of the hour and can be accrued by following the concept of IFS through supplementation of allied agro-enterprises.

The further thrust of IFS is

- There is a need to create the database on farming system in relation to type of farming system, infrastructure, economics, sustainability etc. under different farming situation.
- Need to develop research modules of farming system under different holding size with varying economically viable and socially acceptable systems.
- The assessment and refinement of the technologies developed at research station at cultivators' field.
- Need to prepare a contingent planning to counteract the weather vagaries/ climate threats under different farming situation.
- Need to prepare a policy draft for the consideration of planners for its promotion at large scale with nominal financial assistance either through short/ medium/ long term loans and other promotional advantage.

9

Components of IFS and Its Advantages

Points to be considered while choosing the Enterprises for Integrated Farming System (IFS)

1. Soil and climatic feature of an area/ locality.
2. Resource availability with the farmers.
3. Present level of utilization of resources.
4. Economics of proposed integrated farming system.
5. Farmers managerial skill.
6. Social customs recalling in the locality.

Integration of Enterprises

In agriculture, crop husbandry is the main activity. The income obtained from cropping is hardily sufficient to sustain the farm family throughout year. Activities such as dairying, poultry, fish- culture, sericulture, biogas production, edible mushroom cultivation, agro- forestry, agri- horticulture etc. assume critical importance in supplementing their farm income. Assumed regular cash flow is possible when cropping is combined with other enterprises. Judicious combination of enterprises keeping in view of the environment condition of a locality will pay grater dividends. At the same time it will promote effective recycling of residues/ wastes.

Integrated farming system seems to be the answer to the problems of increasing food production. For increasing income and for improving nutrition of the small scale farmers with limited resources without any adverse effect on environment and agro- eco-system. In a cropping system the amount of by products can be as high as or higher than marketable produce. This may go to waste if not utilized in an animal waste component and to reduce the cost of production of the economic produce of component two and finally to entrance the net income of the farm is whole.

Livestock is the best complementary enterprises with cropping, especially during the adverse years. Installation of a biogas plant in crop- livestock system will make use of the wastes, at the same time provides valuable manure and gas for cooking and lighting.

In a wetland farm, there are greater avenues for fishery, duck farming and buffalo rearing; utilizing the rice straw and mushroom production can be started.

Under irrigated condition (garden lands) inclusion of sericulture, poultry and piggery along with arable crop production is an accepted practice. The poultry component in this system can make use crop the grains produced in the farm as feed. Pigs are the unique components that can be reared with the waste which are unfit for human consumption.

In Rainfed farming, sheep and goat rearing form an integrated part of the landscape, sericulture can be introduced in Rainfed farming provided the climatic condition permit it.

Agro- forestry (silvi- culture and silvi- horticulture) is \the other activities, which can be included under dry land condition. In the integrated system, selection of enterprise should be on the cardinal principal that there should be minimum competition and maximum complementary effect among the enterprises.

1. **Farming System Under Lowland:** Common cropping system in rice based system, especially under agro climatic condition of south India is rice- rice – pulse, modified cropping system includes crops like maze, groundnut, sesamum, rice – fish – poultry culture system appears to more remunerative, poultry droppings from the poultry shed placed well above the farm poned meets the needs of fish in the pond water, Water in the pond can be used for irrigating the crops. About 500 layer chicks are sufficient and excreta can meet the feed requirement of 7, 500 polyculture fingerlings in one ha of pond water rice- fish system is also remunerative.
2. **Upland Irrigated Farming Systems:** Additional income can be generated by enterprises like dairy biogas and silvi culture to the usual cropping systems. Two to three milch cows can be maintained from one ha straw. Recycling of farm and animal wastes through biogas unit can produce cooking gas for family use. Several such integration can greatly increase farm income, besides providing work to family members all through the year.
3. **Rainfed Farming Systems:** Environmentally sustainable dry land farming systems emphasis conservation and utilization of natural resources. Agronomic practices of conservation, tillage and mulch farming, rotational cropping use of legumes and cover crops for improving soil fertility and suppressing weeds and efficient use of cattle manure are some of the components of sustainable farming systems.

For above regulation sound land use policy is necessary to tackle the problems of deteriorating natural resources, like soil and water. Majority farmers maintaining work animals, milch animal, chicks, sheep, goats etc. with crop production. Farmers approaching farming systems, but benefits from these system is low, because of non- educating and also due to non- adoption of improved technology.

Components of Farming Systems

In the integrated farming system, it is always emphasized to combine cropping with other enterprises/ activities, many enterprises are available and these includes cattle maintenance sheep or goat rearing, poultry, piggery, rabbit rearing, bee keeping etc. Any one or more can be combined with the cropping.

Cattle Maintenance

1. Draft breeds
2. Dairy breeds
3. Dual purpose
4. Exotic breeds

1. **Buffaloes:** Important dairy breeds of buffalo are murrah, mehsana, zefarabadi, Godavari.

   Feeding: Cattle feed generally contains fibrous coarse low nutrient straw material. Roughage is basic for cattle ration and includes legumes non- legume hays, straw and silage of legume and grasses. Per day requirement @ 1 kg concentrate per 2 lit of milk, green fodder (20- 30 kg), straw 5-7 kg & water – 32 lit.

2. **Sheep Rearing:** Sheep are well adapted to many areas. They are excellent gleaners and make use of much of waste feed. Consume roughage, converting a relative cheap food into a good cash product. Housing not expensive. Feeding: 1-2 kg of leguminous hay per day. Protein supplied through concentrate as groundnut cake.

3. **Goat Rearing:** In India, activity of goat rearing under different environments. The activity is also associated with different systems such as crop or animal based, single animal or mixed herd small or large scale. Goat is mainly reared foe meat, milk hide and skin meat preferred in India, A goat on hoof fetches a better price than a sheep on hoof.

   Feeding: per head nutrients requirement to goat is low. Hence they are suitable for resources poor small farmers with marginal grazing lands they eat plants and leaves of tree, which any other animals not touch. Goat eats 4- 5 times that of body weight concentrate of maize, groundnut cake etc. and clean and fresh water.

4. **Poultry:** Poultry is one of the fastest growing food industries in the world. Poultry meat accounts for about 27% of total meat consumed worldwide poultry industry in India is relatively a new agricultural industry. Egg production has reached to 5000 crores and broiler meat production 330 thousand tones. The average global consumption is 120 eggs per person/ year and in India it is only 32- 33 eggs per capita/ year. To meet the nutritional requirement the per capita consumption estimated at 180 eggs 9 kg meat/ year.

   Feed: The feed conservation efficiency of the bird is superior to other animals. About 60 – 70 % of the total expenditure on poultry farming is spent on the poultry feed. Hence, use of cheap and efficient ration will give maximum profit cereals- maize, barley, oats, wheat, rice – broken mineral/ salt – limestone, salt manganese.

5. **Duck rearing:** Ducks account for about 7 % of the poultry population in India. They are popular in cereal and logged states like west Bengal. Orissa, A.P, T. Nadu, they have production potential of about 130- 140 eggs/ bird/ year. These can rear in marshy riverside Westland. Duck farming can be a better alternative.

   Feeding: Eating fallen grains in harvested paddy fields, small fishes and other aquatic materials. A variety of crop residues and insects in the farm.

6. **Turkey rearing:** Turkey is a robust bird and can be reared in humid tropics. It actively feed on a variety of crop residues and insects in the farm.

7. **Piggery:** Pigs are maintained for production of pork.

8. **Rabbit Rearing:** In India is of recent origin though hunting of wild rabbits for meat is not uncommon. Rabbit can be easily reared with relatively less concentrate feed with high production rate.

9. **Bee Keeping:** Bee keeping is one of the most important agro- based industries which do not require any raw material like other industries. Nectar and pollen from flower are the raw material, which available in plenty in nature. Bee keeping can be started with a single colony.

   Honey collection: Honey should have good quality. Qualities such as aroma, color, consistency and floral sources are important. Honey is an excellent energy food with an average of about 3500 calories per kg. it is directly absorbed into the human blood stream requiring no digestion.

10. **Aquaculture**: Ponds serve as domestic requirement of water, supplementary irrigation to crop and pisciculture with the traditional management, farmer obtain hardly 300 – 400 of wild and culture fish/ ha/ year. However, polyfish culture with the stocking density of 7500 fingerlings and supplementary feeding will boost the total biomass production.

Species: cattle, Rohu, common carp, silver carp, and grass carp (feed on aquatic plants).

Management: Pond depth – 1.5- 2.0 m, water should be slightly alkaline, PH- 7.5- 8.5. If the PH less than 6.5, it can be adjusted with addition of lime, higher PH (> 8.5) can be reduced with addition of Gypsum. Application of fresh dung may also reduce high PH in the water.

The fish are to be nourished with supplementary feeding with rice bran and oilseed cakes.

This will enable faster growth and better yield. Each variety of crops stocked to 500 fingerlings with the total of 5000- 8000/ ha. This gives 2000 to 5000 kg/ha of fish annually.

11. **Sericulture:** Definition: the keeping of silk moths and their larvae for the production of silk or Seri culture is defined as a practice of combining mulberry cultivation; silkworm rearing and silk reeling. Sericulture is a recognized practice in India. The total area under mulberry is 240 thousand ha in the country. It plays an important role in in socio- economic development of rural poor in some areas. Climatic condition favorable for mulberry and rearing of silk worms throughout the year. Karnataka is the major silk producing state in India (temp 21 to 30 o C), in Kashmir climate suit from May to October.

    Moriculture: Cultivation of mulberry plants is called as Moriculture. The crop yield well for 12 years. Yield of mulberry leaves is 30- 40 t/ ha/ year.

    Rearing: eggs are allowed to be laid over a cardboard. In Bamboo tray rice husk is spread. Tender chopped mulberry leaves are added to the tray. The hatched out larvae are transferred to the leaves it is important to change the leaves every 2 – 3 hours during the first 2 – 3 days. The cocoon constructed with silk. The cocoons required for further rearing are kept separately and moths are allowed to emerge from them.

12. **Mushroom Cultivation:** Mushroom is an edible fungi great diversity in shape, size and colour. Essentially mushroom is a vegetable that is a cultivated that is cultivated in protected farms in a highly sanitized atmosphere; mushroom contains 90 % moisture with in quality protein, fairly good source of vitamin C and B complex. It is rich source of mineral like ca, P, K & Cu. They contain less of fat and CHO and are considered good for diabetic and blood pressure patients.

    Varieties: 1) Oyster mushroom 2) n Paddy straw mushroom- volvarilla volvacea 3) White button mushroom- Agaricus gisporus (var, A-11, Horst V3).

13. **Biogas Plant:** Biogas is a clean, unpolluted and cheap source of energy, which can be obtained by a simple mechanism and little investment. The gas is generated from the cow dung during anaerobic decomposition. Biogas generation is a complex bio- chemical process, celluloytic material are broken down in methane and Co_2 by different group of micro- organism. IT can be used for cooking purpose, burning lamps, etc. Biogas near to kitchen & cattle shed to reduce cost of gas transfer and cow dung transport, sunlight is important for temperature.

Biogas slurry: slurry is obtained after the production of biogas. It is enriched manure; another positive aspect of this manure is that even after weeks of exposure to the atmosphere the slurry does not attract fleas and worms. Dry slurry contains about 1.8 % N, 1.10 % P & 1.50% K.

Benefits or Advantages of Integrated Farming System

1. **Productivity**: IFS provides an opportunity to increase economic yield per unit area per unit time by virtue of intensification of crop and allied enterprises.
2. **Profitability:** Use waste material of one component at the least cost. Thus reduction of cost of production and form the linkage of utilization of waste material, elimination of middleman interference in most input used. Working out net profit B/ C ratio is increased.
3. **Potentiality or Sustainability:** Organic supplementation through effective utilization of by products of linked component is done thus providing an opportunity to sustain the potentiality of production base for much longer periods.
4. **Balanced Food:** We link components of varied nature enabling to produce different sources of nutrition.
5. **Environmental Safety:** In IFS waste materials are effectively recycled by linking appropriate components, thus minimize environment pollution.
6. **Recycling:** Effective recycling of waste material in IFS.
7. **Income Rounds the year:** Due to interaction of enterprises with crops, eggs, milk, mushroom, honey, cocoons silkworm. Provides flow of money to the farmer round the year.
8. **Adoption of New Technology:** Resources farmer (big farmer) fully utilize technology. IFS farmers, linkage of dairy / mushroom / sericulture / vegetable. Money flow round the year gives an inducement to the small/ original farmers to go for the adoption technologies.
9. **Saving Energy:** To identify an alternative source to reduce our dependence on fossil energy source within short time. Effective recycling technique the

organic wastes available in the system can be utilized to generate biogas. Energy crisis can be postponed to the later period.

10. **Meeting Fodder crisis:** Every piece of land area is effectively utilized. Plantation of perennial legume fodder trees on field borders and also fixing the atmospheric nitrogen. These practices will greatly relieve the problem of non – availability of quality fodder to the animal component linked.
11. **Solving Fuel and Timber Crisis:** Linking agro- forestry appropriately the production level of fuel and industrial wood can be enhanced without determining effect on crop. This will also greatly reduce deforestation, preserving our natural ecosystem.
12. **Employment Generation:** Combing crop with livestock enterprises would increase the labour requirement significantly and would help in reducing the problems of under employment to a great extent IFS provide enough scope to employ family labour round the year.
13. **Agro – industries:** When one of produce linked in IFS are increased to commercial level there is surplus value adoption leading to development of allied agro – industries.
14. **Increasing Input Efficiency:** IFS provide good scope to use inputs in different component greater efficiency and benefit cost ratio.

10

Site Specific Development of IFS Model for Different Agro-Climatic Zones

The classification of the farming systems of developing regions may be based on the following criteria:

- Available natural resource base, including water, land, grazing areas and forest; climate, of which altitude is one important determinant; landscape, including slope; farm size, tenure and organization; and
- Dominant pattern of farm activities and household livelihoods, including field crops, livestock, trees, aquaculture, hunting and gathering, processing and off-farm activities; and taking into account the main technologies used, which determine the intensity of production and integration of crops, livestock and other activities.

Based on these criteria, the following eight broad categories of farming system have been distinguished:

- Irrigated farming systems, embracing a broad range of food and cash crop production;
- Wetland rice based farming systems, dependent upon monsoon rains supplemented by irrigation;
- Rainfed farming systems in humid areas of high resource potential, characterized by a crop activity (notably root crops, cereals, industrial tree crops - both small scale and plantation - and commercial horticulture) or mixed crop-livestock systems;
- Rainfed farming systems in steep and highland areas, which are often mixed crop-livestock systems;
- Rainfed farming systems in dry or cold low potential areas, with mixed crop-livestock and pastoral systems merging into sparse and often dispersed systems with very low current productivity or potential because of extreme aridity or cold;
- Dualistic (mixed large commercial and small holder) farming systems, across a variety of ecologies and with diverse production patterns;

- Coastal artisanal fishing, often mixed farming systems; and
- Urban based farming systems, typically focused on horticultural and livestock production.

IFS models for wetland, rainfed and dryland situations

Wetland system

Integrated farming system components have been tested in Tamil Nadu by the scientists of TNAU for wet land condition at Coimbatore (Rangasmy et al. 1996) and at Cauvery Delta zone, Aduthurai (Govindan et al. 1990). Details of various components are given below:

Components of integrated farming for wetland (Coimbatore) are as follows:

1. Cropping
2. Fish culture
3. Poultry
4. Mushroom production

Allocation of area (Total 0.4 ha):

Cropping: 0.36 ha

Fish pond: 0.04 ha

Cropping	Area (ha)
Rice-Rice-Maize	0.16
Rice-Rice-Groundnut	0.10
Maize-Rice-Sesame	0.36

This farming system was compared with the conventional cropping normally followed in the region: rice-rice-green gram (0.20 ha) and rice-rice-green manure (0.20 ha).

Economic returns from the system

On an average, a net profit of Rs. 11,755 was obtained in rice-poultry-mushroom system as compared to Rs. 6335 only from conventional system of cropping alone. An additional employment of 174 man days was generated due to the adoption of IFS.

Integrated farming system-Cauvery Delta zone, Aduthurai, Tamil Nadu

Farmers of this zone are practicing mono-cropping of rice for two seasons (June October and October-January) followed by a rice fallow pulse (January-March). Among the different allied activities, pisci-culture plays an important role in this zone since water is available in the canal for about 7-8 months. Poultry farming is another feasible enterprise. By combing the enterprises of poultry-cum-fish culture with rice cropping system, the economic status of the small and marginal farmers could be improved.

Components description

An area of one hectare has been selected for the study. About 0.04 ha was allotted for fish pond. The improved cropping system of rice- rice -cotton (0.76 ha) and rice-rice- maize (0.20, ha) was adopted. Maize, being a major constituent of poultry feed, was included in the system. This integrated system was compared with the existing practice of rice-rice- black gram. Poultry and fish were the major components of this system.

Economic returns from the system

A total of 4670 eggs were obtained in 22 weeks from 50 birds. In addition, one bird on an average voided about 110 g of droppings per day (wet drops). About 794 kg of fresh droppings (260 kg dry) were recycled in the fish pond during the period of 22 weeks. The growth rate of fish was linear for each month of sampling. During the harvest at the end of 6 1/2 months, the total fish weight was 450 kg/ ha in the treatment pond and 240 kg/ha in the control pond. The increased fish production in the treatment pond was due to recycling of 350 kg poultry droppings as feed.

A net return of Rs. 17,200 was obtained by integrating different enterprises. By introducing poultry-cum-fish culture with cropping, a total employment of 385 man-days was generated.

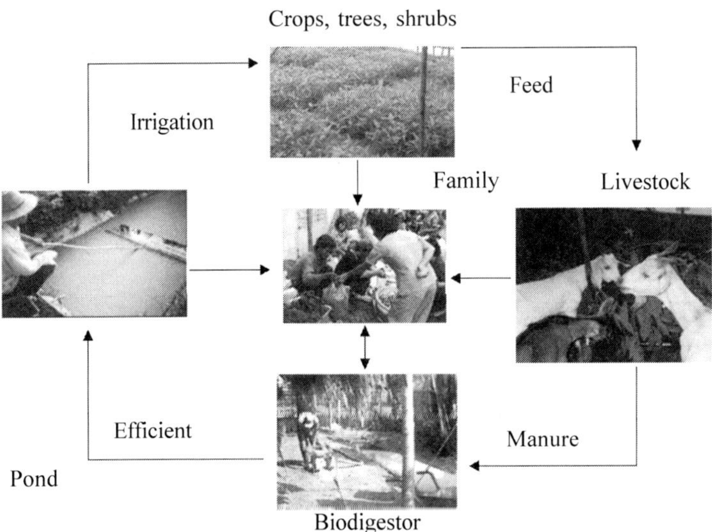

System for irrigated situations (Garden lands)

A model integrated farming system to suit the small and marginal famers of garden land conditions was studied at TNAU, Coimbatore (Rangaswamy et al. 1996)., during 1988-1993. An area of one ha was selected for IFS and compared with conventional cropping system (CCS).

Components of IFS

S. No.	Cropping	Area (ha)
1	Cotton + green gram-maize +fodder cowpea-bellary onion	0.56
2	Wheat + sunflower – maize + fodder cowpea – summer cotton + green gram	0.19
3	Grass Bajra Napier (CO.I)	0.15
4	Lucerne	0.05
5	150 trees of *Leuceana* (planted in the bunds)	0.05
	Total	1.00

Farmstead

Dairy unit	3 Jersey cows + 2 calves
Biogas unit	2 m^3 capacity
Mushroom production	1.5-2.0kg/day

The above integrated system was compared with the conventional cropping system of cotton-sorghum-finger millet in 0.20 ha area.

C: CROP - FISH – POULTRY FARMING SYSTEM

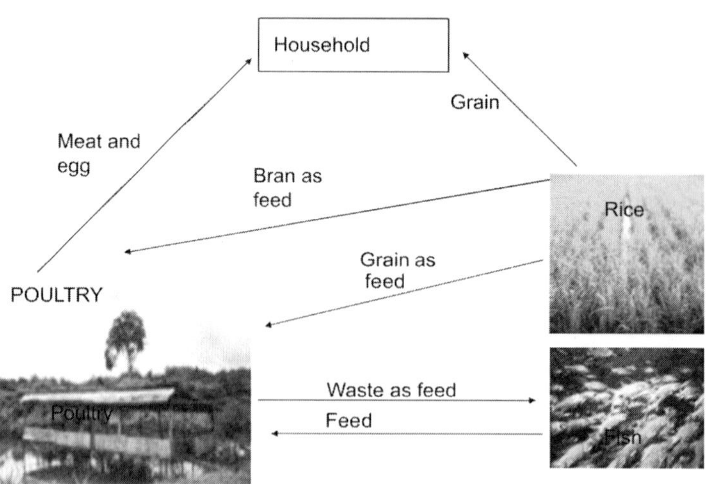

Economic returns from the system

Maize flour, cottonseed and wheat bran obtained from the crop components were recycled for preparing dairy from the second year. About 45.5 t of grass fodder, 2.5 t of legume fodder and 1.0 t of dry fodder were obtained from the system and fed to the animals. Dung was recycled for the biogas plant. Mean revenue of Rs. 34,600/ha was realized in IFS as compared to Rs. 13,950 obtained in CCS. Employment opportunity was also enhanced to the tune of 770 man days per year under IFS as against conventional cropping.

Dryland based system

Integrated farming system for dryland suggested for Coimbatore and Aruppukottai, Tamil Nadu, are described below:

Model for Coimbatore, Tamil Nadu

(Mean annual rainfall: 640 mm)

Crop components for one hectare

S. No.	Cropping	Area (ha)
1	Sorghum + cowpea both grain purpose	0.20
2	Sorghum + cowpea both fodder purpose	0.20
3	*Leucaena* (tree fodder) + *Cenchrus* (Grass fooder)	0.20
4	*Acacia Senegal* (tree fodder) + grass	0.20
5	*Prosonic cineraris* (tree fodder) + grass	0.20

Animal components: Tellicherry goat: 6(5 female + 1 male-stall fed)

Conventional cropping system = 0.20 ha

(Sorghum + cowpea-gain purpose)

Economic returns from the system: Mean additional revenue of Rs 3750/ha/ year was obtained from IFS over CCS. The employment generation under IFS was 153 man days/ha/year whereas it was only 40 man days/ha/year in the CCS (Sivasankaran et al. 1995).

Model for Arupukottai, Tamil Nadu

Integrated farming system study for drylands was conducted from 1989-90 to 1994-95 at Regional Research Station, Arupukottai, Tamil Nadu. The soil type of the experimental site was black cotton soil (vertisol) and the annual rainfall of the region is 830 mm received in 47 rainy days. The rainfall pattern is "unimodel' with its peak during the months of October and November. The following is the integrated farming system model that has been studied (cropping – 1.6 ha, fruit trees-0.4 ha). This was compared with conventional system of sorghum alone.

Economics of the system

The additional net income obtained from IFS over CCS varied from Rs. 2,160 to 15,460 in different years. An increase of additional net income was noticed during each year and it was higher during the sixth year. A mean additional net income of Rs. 7100 per year was obtained in IFS over the CCS. The additional net income per day in IFS over CCS was Rs. 2160. Employment generation under IFS was 131 man days per ha per year, whereas it was only 35 man days per ha per year under CCS, thus generating an additional employment of 78 man days/ha/year under IFS.

Cropping	Area (ha)
Cotton + black gram	0.5
Sorghum + cowpea	0.5
Fodder crops (*Cenchrus ciliaris*)	0.2
Fodder tress	0.4
Fruit tress	
Ber	0.2
Custard apple	0.1
Aonla	0.1

D: Crop (livestock-poultry-fishery farming system)

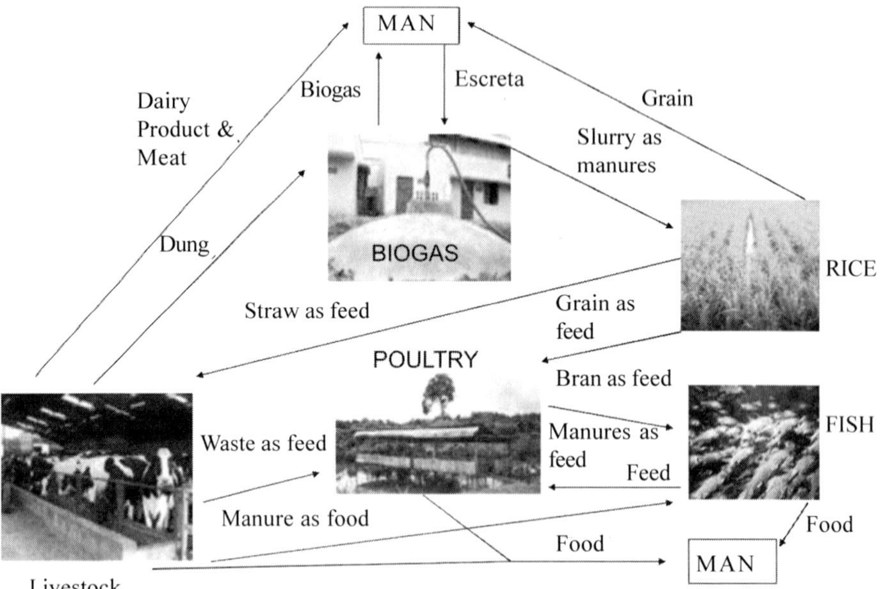

Suggested cropping systems / farming systems under different farming situations in Haryana.

S.No.	Farming situations	Cropping systems	Farming systems
1.	Shivalik foot hills (high rainfall i.e. 1000 mm and above)		
	a) After developing water sheds.	Arhar-wheat	Mixed farming with three cross-bred cows+crops
	b) Non-watershed area	Maize+urd/soybean fallow Groundnut-fallow Soybean-field pea/mustard Arhar-fallow	20 sheep/goats farming+crops
2.	Plain irrigated area with good quality under ground water	Paddy-wheat-green manuring	Three cross bred cows/three
		Paddy-wheat/sorghum-wheat/Paddy-wheat/sugarcane-ratoon	Buffaloes + crops
		Paddy-potato-onion/bhindi	Vegetable/Horticulture near cities
3.	Plain irrigated and brackish ground water	Sorghum-wheat/sugarcane-ratoon Sorghum-wheat/sorghum-mustard or cotton-wheat/ Arhar-wheat/ Sorghum-berseem	Mixed farming with two crossbred cows/two buffaloes
4.	Limited irrigation with brackish water	Bajra-Mustard Bajra-wheat Clusterbean-wheat Sorghum-Oats	Mixed farming with two buffaloes+crops
5.	Rainfed in low rainfall area (400-800 mm)	Bajra-gram Arhar+moong-fallow Fallow-mustard/gram Clusterbean-fallow/bajra-gram	Mixed farming with one buffalo+crops
6.	Rainfed dunal area	Caster+Moong/Clusterbean Bajra-Fallow/Clusterbean-Fallow/ Fallow-gram	Silvipasture system with Goats/Sheep(20animals)

11

Resource Use Efficiency & Optimization Techniques; Resource Cycling and Flow of Energy in Different Farming System; Farming System and Environment

Resource use efficiency in Indian agriculture

Resource use efficiency in agriculture is defined to include the concepts of technical efficiency, allocative efficiency and environmental efficiency. Public investment, subsidies and credit for agriculture are used in an efficient manner. There are large scale inter regional as well as inter farm variations in factor productivity due to varying influence of different factors in different regions. A number of management factors such as timeliness and method of sowing, transplanting, irrigation and application of right doses of inputs and input mix play an important role in influencing inter-farm variation in crop productivity. Growing marginalization and fragmentation of land holdings coupled with rising incidence of informal tenancies and poor rural infrastructure such as road, electricity, markets and education affect factor productivity. The availability of good quality irrigation water coupled with flexibility of irrigation and drainage system and appropriate methods of application as well as pricing of irrigation water is crucial for sustainable use of land and water resources.

The stochastic frontier production function to analyse the resource use efficiency of urban farmers in Uyo, Southeastern Nigeria was done. The result shows that 65% of urban farmers were 70% technology efficient; maximum efficiency is 0.91, while minimum efficiency in urban farm is 0.43.

Three types of efficiency identified in the literature viz. technical efficiency, allocative efficiency and overall or economic efficiency are defied as; Technical efficiency is the ability of a firm to produce a given level of output with minimum quantity of inputs under a given technology. Allocative efficiency is a measure of the degree of success in achieving the best combination of different inputs in producing a specific level of output considering the relative prices of these inputs.

Economic efficiency is a product of technical and allocative efficiency. In one sense, the efficiency of a firm is its success in producing as large an amount of output as possible from given sets of inputs. Maximum efficiency of a firm is attained when it becomes impossible to reshuffle a given resource combination without decreasing the total output.

These studies have employed several measures of efficiency. These measures have been classified broadly into three namely: deterministic parametric estimation, non- parametric mathematical programming and the stochastic parametric estimation. There are two non-parametric measures of efficiency. The first, based on the work of many workers evaluates efficiency based on the neoclassical theories of consistency, restriction of production form, recoverability and extrapolation without maintaining any hypothesis of functional form.

Several approaches, which fall under the two broad groups of parametric and non-parametric methods, have been used in empirical studies of farm efficiency. These include the production functions, programming techniques and recently, the efficiency frontier. The frontier is concerned with the concept of maximality in which the function sets a limit to the range of possible observations. Thus, it is possible to observe points below the production frontier for firms producing less than the maximum possible output but no point can lie above the production frontier, given the technology available. The frontier represents an efficient technology and deviation from the frontier is regarded as inefficient.

The literature emphasizes two broad approaches to production frontier estimation and technical efficiency measurement: (a) The non-parametric programming approach, and (b) the statistical approach. The programming approach requires the construction of a free disposal convex hull in the input-output space from a given sample of observations of inputs and outputs. The convex hull (generated from a subset of the given sample) serves as an estimate of the production frontier, depicting the maximum possible output. Production efficiency of an economic unit is thus measured as the ratio of the actual output to the maximum output possible on the convex hull corresponding to the given set of inputs.

Resource cycling and flow of energy in different farming system

Recycling of organic wastes such as crop residues, dung and urine from domesticated animals and wastage from slaughter house, human excreta and sewage, bio mass of weeds, organic wastes from fruit and vegetables production and household wastes, sugarcane trash, oilcakes, press mud and fly ash from thermal power plants is the fundamental in the sustainability of farming system. Material not suitable for direct application can be applied after composting or vermicomposting.

The ultimate goal of sustainable agriculture is to develop farming systems that are productive and profitable, conserve the natural resources base, protect the environment and enhance health and safety, and to do so over the long term.

Two farming system have been proposed for ensuring sustainability. There are low input sustainable agriculture (LISA) and organic farming.

Low Input Sustainable Agriculture (LISA):

In this system minimal use of external production inputs is made. The production costs are obviously lower. The overall risk of the farmer is considerably reduced. Besides, the above advantages, pollution of surface and ground water is avoided and healthy food with very little or no pesticide residues is ensured. These systems held promise for both short and long term profitability. However, the system suffers from one serious drawback – continuation of low external input agriculture will perpetuate to the vicious circle of low inputs low yields which the third world countries with ever increasing population pressure can ill offered.

High input system, on the other hand, will fail sooner or later, as they are not economically and environmentally sustainable. What is the solution then? The optimal input farming system has the premise of low input per unit of output and lays emphasis on law of diminishing returns. Four factors are most important for better growth of plant:

1. Soil – 45%
2. Organic matter – 5%
3. Air – 25%
4. Water = 25%

Farming system and environment

All over the world, farmers work hard but do not make money, especially small farmers because there is very little left after they pay for all inputs (seeds, livestock breeds, fertilizers, pesticides, energy, feed, labour, etc.). The emergence of Integrated Farming Systems (IFS) has enabled us to develop a framework for an alternative development model to improve the feasibility of small sized farming operations in relation to larger ones. Integrated farming system (or integrated agriculture) is a commonly and broadly used word to explain a more integrated approach to farming as compared to monoculture approaches. It refers to agricultural systems that integrate livestock and crop production or integrate fish and livestock and may sometimes be known as Integrated Biosystems. In this system an inter-related set of enterprises used so that the "waste" from one component becomes an input for another part of the system, which reduces cost and improves production and/or income. IFS works as a system of systems. IFS ensure that wastes from one form of agriculture become a resource for another form. Since it utilizes wastes as resources, we not only eliminate wastes but we also ensure overall increase in productivity for the whole agricultural systems. We avoid the environmental impacts caused by wastes from intensive activities such as pig farming.

Integrated farming systems: Environmental Sustainability in Full Circle

In recent years, food security, livelihood security, water security as well as natural resources conservation and environment protection have emerged as major issues worldwide. Developing countries are struggling to deal with these issues and also have to contend with the dual burden of climate change and globalization.

It has been accepted by decision makers across the globe that sustainable development is the only way to promote rational utilization of resources and environmental protection without hampering economic growth. Different countries around the world are promoting sustainable development through sustainable agricultural practices which will help them in addressing socio-economic as well as environmental issues simultaneously.

Within the broad concept of sustainable agriculture "Integrated Farming Systems" hold special position as in this system nothing is wasted, the byproduct of one system becomes the input for other. Integrated farming is an integrated approach to farming as compared to existing monoculture approaches. It refers to agricultural systems that integrate livestock and crop production. Moreover, the system help poor small farmers, who have very small land holding for crop production and a few heads of livestock to diversify farm production, increase cash income, improve quality and quantity of food produced and exploitation of unutilized resources.

Components of integration in a farming system are parkland systems, trees on bunds, wind breaks, silvi- pasture system, agro-horticulture system, block plantations, economic shrubs, live fences, crops with green leaf manure species (mixed/intercrops), integrated animal based systems (fisheries, dairy, piggery, small ruminants, poultry, apiary).

The intensive farming systems of developed countries, such as United Kingdom seek to maximize yield through what is usually described by agricultural economists as Best Management Practice (BMP), which involves the most efficient use of all inputs, including fertilizers, herbicides, seed varieties, and precision agricultural techniques. Fertilizers have been central to this approach, which has resulted in a tremendous increase in productivity over that last 40 years. For example, the efficient use of improved fertilizers, combined with new varieties of wheat and the successful use of crop protection chemicals, has increased grain yields from 3 tonnes per hectare to approximately 10 to 11 tons per hectare today. Moreover the current market economic incentives facing many farmers are likely to encourage excess fertilizer application. It is generally recognized that if eventually the adoption of market prices for most agricultural goods without any subsidies became a reality, in order to be competitive with the lower production costs of developing countries in South America, Asia, Eastern Europe and the Former Soviet Union, the pressure to intensify even the most United Kingdom intensive production systems will as well become reality despite the negative consequences on the environment.

Agriculture is one of the most successful sectors in terms of productivity growth, has outpaced the rapid growth in demand for its output for the past decades. This trend has provided hefty social benefits, such as increased the accessibility of agricultural goods usually at a lower price, provision of jobs and therefore rural sustainability, energy and also positive environmental effects, such as aesthetic value, carbon sequestration by soils and trees, and other additional benefits that are linked with good husbandry such as maintenance of natural habitats and countryside landscape. However, is largely referenced in literature that the increased use of chemicals either fertilizers or pesticides in agriculture intensive systems is associated with hidden costs due to environmental pollution in soil, water and atmosphere –, consequently has amplified the negative social effects on the natural environment. This argument is supported by an analysis of the externalities from UNITED KINGDOM agriculture that negative externalities amount to at least £1 billion, and positive externalities offset approximately half of these negative effects (negative/positive external).

Integrated crop–livestock systems: Strategies to achieve synergy between agricultural production and environmental quality

A need to increase agricultural production across the world for food security appears to be at odds with the urgency to reduce agriculture's negative environmental impacts. We suggest that a cause of this dichotomy is loss of diversity within agricultural systems at field, farm and landscape scales. To increase diversity, local integration of cropping with livestock systems is suggested, which would allow:

i. Better regulation of biogeochemical cycles and decreased environmental fluxes to the atmosphere and hydrosphere through spatial and temporal interactions among different land-use systems;

ii. A more diversified and structured landscape mosaic that would favour diverse habitats and trophic networks; and

iii. Greater flexibility of the whole system to cope with potential socio-economic and climate change induced hazards and crises.

The fundamental role of grasslands on the reduction of environmental fluxes to the atmosphere and hydrosphere operates through the coupling of C and N cycles within vegetation, soil organic matter and soil microbial biomass. Therefore, close association of grassland systems with cropping systems should help mitigate negative environmental impacts resulting from intensification of cropping systems and improve the quality of grasslands through periodic renovations. However, much research is needed on designing appropriate spatial and temporal interactions between these systems using contemporary technologies to achieve the greatest benefits in different agro-ecological regions. We postulate that development of modern integrated crop–livestock systems to increase food production at farm and regional levels could be achieved, while improving many ecosystem services. Integrated crop–livestock systems, therefore, could be a key form of ecological intensification needed for achieving future food security and environmental sustainability.

Optimizing soil, water and nutrient use efficiency in integrated crop-livestock farming systems

The introduction, adaptation and implementation of good farming practices are needed to address the challenges of increasing global population, food scarcity and making agriculture resilient to climate change. These practices are also important to maintain local and national food security and livelihoods improve agricultural resource use efficiency and provide social and economic benefits. The integration of crops with livestock in agricultural production systems is a winning combination which results in many benefits such as:

1. Conserving natural resources,
2. Enhancing ecosystem services and environmental sustainability,
3. Providing natural pest control
4. Improving soil quality and crop yield, and
5. Reducing risk through diversification of crop and livestock enterprises.

However, many of these benefits have not yet been quantified or clearly understood. Persuading farmers to adopt integrated crop-livestock production systems, and policy makers to provide institutional support for implementing these systems is critically important. Success depends on the provision of quantitative information on the economic, environment and resource use benefits of such systems. The proposed CRP assesses farming practices to improve the efficiency of soil, water and nutrient use in integrated crop-livestock production systems and to unlock the resources for improving food security. Both outcomes depend on identifying and quantifying improved management of soil, water, nutrient and soil quality, GHG emissions and nutrient losses from animal manure and crop residues. Using isotopic and nuclear techniques will provide more insight on the soil-water-crop-nutrient-animal interactions in integrated crop-livestock production systems.

Objectives

- To enhance food security and rural livelihoods by improving resource use efficiency and sustainability of integrated crop-livestock systems under a changing climate
- To assess socio-economic and environmental benefits of crop-livestock systems
- To assess the influence of crop - livestock systems on GHG emissions, soil carbon sequestration and water quality
- To develop soil, water and nutrient management options in integrated crop-livestock systems for potential adoption by farmers
- To identify the potential for improving soil quality
- To optimize water and nutrient use efficiency in integrated crop-livestock production systems
- To strengthen the capacity of member states to use isotopic and nuclear techniques as tools for improving the management of crop-livestock systems

Resource Flow in integrated farming systems

Conceptually, IFS involves a blend of inter alia crops, animals, poultry, pisciculture, bee keeping, sericulture, mushroom cultivation, and agroforestry. The concept of conservation and recycling of energy and mineral matter in the soil-plant-animal-atmosphere continuum is followed. A farming system in much of the developed world has become an issue of managing a set of individual enterprises. Individual farm enterprises driven by advancing technology have developed almost in isolation. Industrial inputs into farming have almost broken the subsystem (enterprise) interactions. Certain dependencies between enterprises of course remain; these are related to the need to distribute scarce resources within the farm business. The management of the farm as a system has been relegated to resource acquisition and an allocation problem between (almost) independent enterprises. Farming systems in India and other developing countries are mostly subsistence in nature. The enterprises exist mainly in natural form based on their complementarity. For example, in a fish-duck farming system many complementarities are observed between the duck enterprise and the pond ecosystem. The on-station study conducted AICRP on IFS involving enterprises such as crop, poultry, a fishery, a duckery, an apiary, and mushroom production revealed that there is chain of interactions amongst these enterprises. The by-product of one enterprise may be effectively utilized by the other enterprises, thus ensuring higher and efficient resource use efficiency. A close examination of resource recycling (Fig.) indicates the interdependence of the different components of the total farming system to make the farmer self-sufficient in terms of ensuring family members a balanced diet for leading a healthy life and also making the farm self-sufficient through recycling of by-products and wastes. The by-product of dairying (cow dung) forms a major raw material for bio-gas plants. Digested slurry from bio-gas production forms a useful feed for pisciculture by increasing plankton growth, as well as supplying valuable manure to raise the productivity of field crops and enrich the soil. By-products of field crops such as paddy straw form a major raw material for mushroom cultivation. Straw after use in mushroom production is utilized as cattle feed and compost preparation. Similarly, poultry droppings form an important ingredient of pisciculture for increasing plankton growth as well as increasing land fertility. An apiary, apart from providing a wholesome food product such as honey, plays a role in improving pollination. Therefore, it is dangerous to deal separately with things in such linked system. The entire philosophy of IFS revolves round better utilization of time, money, resources, and family labor. The farm family has scope for gainful employment year round, thereby ensuring a reasonable income and a better standard of living.

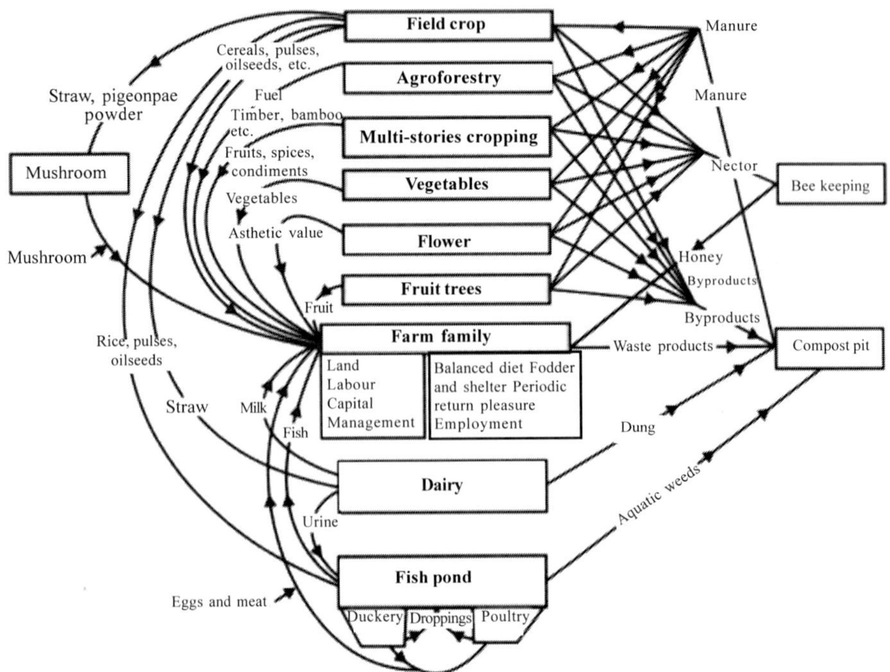

Design of sustainable farming system

Most people involved in the promotion of sustainable agriculture aim at creating a form of agriculture that maintains productivity in the long term by optimizing the use of locally available resources by combining the different components of the farm system, i.e. plants, animals, soil, water, climate and people, so that they complement each other and have the greatest possible synergetic effects.

i) Reducing the use of off-farm, external and non-renewable inputs with the greatest potential to damage the environment or harm the health of farmers and consumers, and a more targeted use of the remaining inputs used with a view to minimizing variable costs.

ii) Relying mainly on resources within the agro ecosystem by replacing external inputs with nutrient cycling, better conservation, and an expanded use of local resources.

iii) Improving the match between cropping patterns and the productive potential and environmental constraints of climate and landscape to ensure long-term sustainability of current production levels.

iv) Working to value and conserve biological diversity, both in the wild and in domesticated landscapes, and making optimal use of the biological and genetic potential of plant and animal species.

v) Taking full advantage of local knowledge and practices, including innovative approaches not yet fully understood by scientists although widely adopted by farmers.

Sustainable farming system provides the knowledge and methodology necessary for developing an agriculture that is on the one hand environmentally sound and on the other hand highly productive, socially equitable and 353 economically viable. Through the application of agro-ecological principles, the basic challenge for sustainable agriculture to make better use of internal resources can be easily achieved by minimizing the external inputs used, and preferably by regenerating internal resources more effectively through diversification strategies that enhance synergisms among key components of the agro-ecosystem.

The ultimate goal of sustainable farming system designs is to integrate components so that overall biological efficiency is improved, biodiversity is preserved, and the agro-ecosystem productivity and its self-regulating capacity are maintained. The goal is to design sustainable farming system that mimics the structure and function of local natural ecosystems; that is, a system with high species diversity and a biologically active soil, one that promotes natural pest control, nutrient recycling and high soil -cover to prevent resource losses.

Sustainable farming system provides guidelines to develop diversified agro ecosystems that take advantage of the effects of the integration of plant and animal biodiversity such integration enhances complex interactions and synergisms and optimizes ecosystem functions and processes, such as biotic regulation of harmful organisms, nutrient recycling, and biomass production and accumulation, thus allowing agro-ecosystems to sponsor their own functioning. The end result of sustainable farming system design is improved economic and ecological sustainability of the agro-ecosystem, with the proposed management systems specifically in tune with the local resource base and operational framework of existing environmental and socioeconomic conditions. In an agro-ecological strategy, management components are directed to highlight the conservation and enhancement of local agricultural resources (germplasm, soil, beneficial fauna, plant biodiversity, *etc.*) by emphasizing a development methodology that encourages farmer participation, use of traditional knowledge, and adaptation of farm enterprises that fit local needs and socioeconomic and biophysical conditions.

Resource(s) use efficiencies and optimization techniques

Resources: land, water, nutrients, energy, labour and capital

Efficiency: Efficiency in any system is an expression of obtainable output with the addition of unit amount of input.

The ratio of energy intake and energy of the produced biomass i.e. of input and output is called ecological efficiency. This can be studied at any trophic level.

Resource use efficiency (fertilizer, water etc) is the output of any crop or anything else per unit of the resource applied under a specified set of soil and climatic conditions.

Examples

Cultivated land utilization Index: is calculated by summing the products of land area to each crop, multiplied by the actual duration of that crop divided by the total cultivated land times 365 days.

$$\text{CLUI} = \frac{\sum_{t=1}^{n} \text{aidi}}{\text{A} \times 365}$$

Where,

n = total number of crops;

a_i = area occupied by the ith crop,

d_i = days that the ith crop occupied and

A = total cultivated land area available for 365 days.

CLUI can be expressed as a fraction or percentage. This gives an idea about how the land area has been put into use. If the index is 1 (100%), it shows that the land has been left fallow and more than 1, tells the specification of intercropping and relay cropping. limitation of CLUI is its inability to consider the land temporarily available to the farmer for cultivation.

Fertilizer use efficiency (FUE): FUE is the output of any crop per unit of the nutrient applied under a specified set of soil and climatic conditions. The NUE/ FUE can be expressed in several ways. Mosier *et al.* (2004) described 4 agronomic indices to describe NUE:

Partial factor productivity (PFP, kg crop yield per kg input applied); **agronomic efficiency** (AE, kg crop yield increase per kg nutrient applied); **apparent recovery efficiency** (RE, kg nutrient taken up per kg nutrient applied); and **physiological efficiency** (PE, kg yield increase per kg nutrient taken up). **Crop removal efficiency** (removal of nutrient in harvested crop as % of nutrient applied) is also commonly used to explain nutrient efficiency.

Energy efficiency: Energy output (Mj/ha)/Energy input (Mj/ha)

Net energy (Mj/ha) = Energy output (Mj/ha) - Energy input (Mj/ha)

Energy productivity (kg/Mj) – Output (grain + byproduct, kg/ha)/Energy input (Mj/ha)

Energy intensity (in physical term, Mj/ha) = Energy output (Mj/ha)/ Output (grain + byproduct, kg/ha)

Energy intensity (in economic terms, Mj/Rs.) = Energy output (Mj/ha)/cost of cultivation (Rs./ha)

Water use efficiency

Crop water use efficiency Crop water use efficiency is a ratio between marketable crop yield and water used by the crop in evapotranspiration.

WUE (kg/ha-mm) = Y/ETc

Where,

WUE = Water use efficiency in kg/ha-mm

Y = Marketable crop yield in kg/ha

ETc = Crop evapotranspiration in mm

Field water use efficiency: Field water use efficiency is a ratio between marketable crop yield and field water supply which includes water used by the plant in metabolic activities, ET and deep percolation losses.

WUE (kg/ha-mm) = Y/WR

Where,

FWUE = Field water use efficiency in kg/ha-mm

Y = Crop yield in kg/ha

WR = Water used in metabolic activities,

ET and deep percolation losses in mm

Water Use Efficiency of Crops

CROP	Water requirement (mm)	Grain yield (kg/ha)	WUE (kg/hamm)
Rice	1200	4500	3.7
Sorghum	500	4500	9.0
Pearl millet	500	4000	8.0
Maize	625	5000	8.0
Groundnut	506	4616	9.2
Wheat	280	3534	12.6
Finger millet	310	4137	13.7

Weed control efficiency: [Weed count or dry weight in weedy - weed count or dry weight in a treatment]/weed count or dry weight in weedy

Resource use efficiency: Resource use efficiency in agriculture is defined to include the concepts of technical efficiency, allocative efficiency and environmental efficiency.

Public investment, subsidies and credit for agriculture are used in an efficient manner. There are large scale inter regional as well as inter farm variations in factor productivity due to varying influence of different factors in different regions. influence of different factors in different regions.

A number of management factors such as timeliness and method of sowing, transplanting, irrigation and application of right doses of inputs and input mix play an important role in influencing inter-farm variation in crop productivity.

Growing marginalization and fragmentation of land holdings coupled with rising incidence of informal tenancies and poor rural infrastructure such as road, electricity, markets and education affect factor productivity.

The availability of good quality irrigation water coupled with flexibility of irrigation and drainage system and appropriate methods of application as well as pricing of irrigation water is crucial for sustainable use of land and water resources.

Technical efficiency is the ability of a firm to produce a given level of output with minimum quantity of inputs under a given technology. Allocative efficiency is a measure of the Allocative efficiency is a measure of the degree of success in achieving the best combination of different inputs in producing a specific level of output considering the relative prices of these inputs.

Economic efficiency is a product of technical and allocative efficiency. In one sense, the efficiency of a firm is its success in producing as large an amount of output as possible from given sets of inputs. Maximum efficiency of a firm is attained when it becomes impossible to reshuffle a given resource becomes impossible to reshuffle a given resource combination without decreasing the total output.

Linear programming: The word linear is used to describe the relationship among two or more variables which are directly proportional. For example, doubling (or tripling) the production of a product will exactly double (or triple) the profit and the required double (or triple) the profit and the required resources, then it is linear relationship.

Programming implies planning of activities in a manner that achieves some optimal result with restricted resources.

Linear programming was developed by George B Dantzing (1947) during Second World War. It has been widely used to find the optimum. It has been widely used to find the optimum resource allocation and enterprise combination. Linear programming is defined as the optimization (Minimization or maximization) of a linear function subject to specific linear inequalities or equalities.

Maximize z $\sum_{i=1}^{n} CjXj$

Such that $\sum_{j=1}^{n} Cij<bi \; = 1 \text{to } m$

Xj = 0
Cj = Net income from jth activity xi = Level of jth activity
Aij = Amount of ith resourcerequired for jth activity
Bi = Amount of ith resource available.

Assumptions of Linear Programming

Linearity: It describes the relationship among two or more variables which are directly proportional.

Additivity: Total input required is the sum of the resources used by each activity. Total product is sum of the production from each activity.

Divisibility: Resources can be used in fractional amounts. Similarly, the output can be produced in fractions.

Finiteness of activities and resource restrictions: There is limit to the number of activities and resource constraints.

Non negativity: Resources and activities cannot take negative values. That means the level of activities or resources cannot be less than zero.

Single value expectations: Resource supplies, input- output coefficients and prices are known with certainty.

Advantages of L.P

i) Allocation problems are solved.

ii) Provides possible and practical solutions.

iii) Improves the quality of decisions.

iv) Highlights the constraints in the production.

v) Helps in optimum use of resources.

vi) Provides information on marginal value products (shadow prices).

Limitations

i) Linearity

ii) Considers only one objective for optimization.

iii) Does not consider the effect of time and uncertainty

iv) No guarantee of integer solutions

v) Single valued expectations.

12

New Concept and Recent Approaches in Farming Systems

Pre-dominant farming systems in various regions of India

The quick survey conducted as a part of characterization of existing farming systems throughout the country indicates existence of 19 pre-dominant farming systems with majority as crop + livestock (85%). Although crop + livestock system is dominating in the country, based on the per cent contribution to net income, the systems are classified as crop, horticulture, livestock, fisheries dominant systems where in dominant component contributes more than 50% of the total net returns. Accordingly, it was found that crop dominant farming systems are existing in the states of Andhra Pradesh, Bihar, Chhattisgarh, Goa, Haryana, Jammu and Kashmir, Jharkhand, Kerala, Karnataka, Madhya Pradesh, North-East, Maharashtra, Orissa, Punjab, Tamil Nadu, Uttar Pradesh and Uttaranchal while livestock dominant systems are present in Rajasthan and Parts of Gujarat. West Bengal, parts of Odisha and Assam states have the fisheries as a major source of income to the existing farming systems. The scope for promotion of horticulture (fruit) based systems exists in Jammu and Kashmir, Himachal Pradesh, Maharashtra, parts of Uttar Pradesh and in Sikkim while plantation dominant systems are available in Andaman and Nicobar Islands and Kerala. In selected states and locations, highly diversified systems also exists where in none of the component contributes for 50% or more to the returns. The characterization survey data of existing farming systems in different states/zones are being interpreted and brought out in the form of National Atlas on Farming Systems. Though the various farming systems exists in the country, integration of input and output within the system is either completely lacking or at partial integration. Competition exists within and outside the farm for various byproducts generated. Cow dung is the best example as dung is required for improving the fertility of soil and meeting the household fuel. Hence, synergy needs to be made by appropriate allocation for making farming as a profitable option. Sustainable farming systems should aim for long term productivity, profitability, recycling of resources and employment generation. The monetary returns under different farming systems practiced in different parts of the country are given in Fig 1. Among the various existing systems in the country, coconut +banana+ cocoa +pineapple + nutmeg (FS9) gives the higher return of Rs 1.27 lakhs/annum.

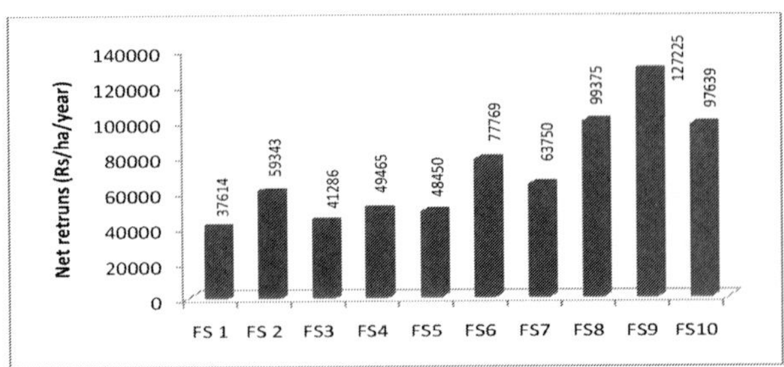

FS 1	Crops + dairy	FS6	Crops+dairy+ horticulture +poultry+ fish
FS 2	Crops +dairy+horticulture	FS7	Rice+fish
FS3	Crops+dairy+poultry	FS8	Coconut +banana
FS4	Crops+dairy+goat	FS9	Coconut+banana+ cocoa+ pineapple + nutmeg
FS5	Crops+fishery+dairy+poultry	FS10	Crop +dairy+horticulture+ apiary+ fish

Fig 1 : Monetary returns of different farming systems practiced in the country (As reported by PDFSR, Modipuram)

Production potential of different components of farming systems; interaction and mechanism of different production factors.

Farming System and Its Characteristics

Farming systems research (FSR)

Originates from recognizing the inter- dependence and inter relationships of natural environment within the farming system. In FSR the farmers by participating the research process help in the identification of the research problems as well as take part in testing the possible solution.

Goals

The growing concern on suitable development has led the FSR to compasses on sound management of farm resources to enhance farm productivity and reduce the degradation of environment quality or to develop sustainable land use, which will optimize farm resource, minimum degradation with consideration to regenerative capacity, increase income and employment for farm families and promote quality of life.

Principal Involved: The FSR approach includes:

a) Viewing the farm as a whole

b) Identifying the farming system, the interacting component and delineating boundaries

c) Systematically investigation the nature and extent of interdependence among the enterprises and identifying constraint

d) Applying the modern technical know- how to the system so as to make it yield optimum results

e) Studying the equity gender income, employment and resources use efficiency

f) Dealing with the issue at integration level through analysis and solution of problems towards sustainable farming system development

In the past decade, farming system research has emerged as a popular and major theme in international agricultural research. FSR evolved in post- green Revolution era with the growing perception of the failure of main stream agricultural research and extension institution to generate and disseminate technologies widely adopted by small scale, resources poor farmers. Clearly technology, even when sound by scientific standards, is of limited value if is not adopted. The diagnosis of the problems was that agricultural researches and development planners, the generators and disseminators of new technology, had employed a fundamentally top- down approach to technology development, which is not valid one. In response to this situation, FSR argued that:

1. Development of relevant and viable technology for small farmers must be grounded in a full knowledge of existing of the farming system and

2. Technology should be evaluated not solely in terms of its technical performance, but in terms of its conformity to the goals, needs and socio-economic condition of small farm system as well. Therefore, FSR concept was developed in 1970 in response to the observation that groups of small – scale farm families operation on harsh environment were not benefiting from conventional agricultural research and extension strategies. The tem FSR in its broadest sense is any research that views the farm in a holistic manner and considers interactions (between component and of components with environment) in the system.

Farming system research is a research method designated to understand farmer's priorities, strategies and resources allocation decisions. It is most often used in conjunction with on farm highly location specific research with multi and inter disciplinary in nature and uses a whole farm approach for improved technologies to enhance and stabilize agriculture production.

The Characteristics of Farming System Research

1. It is holistic or system oriented

2. It is problems solving: involvement of farmers in problem identification and solving

3. It is farmer participatory
4. It envisages location specific technology solutions
5. It is for specific client group – small/ marginal farmer
6. It adopts bottom up approach
7. It compasses extensive on farm activities, collaboration between farmer and scientist
8. It is gender sensitive
9. It ultimate objective is sustainability
10. It focuses on actual adoption
11. It recognizes interdependence among multiple clients

Points to be considered while choosing the Enterprises for Integrated Farming System (IFS)

1. Soil and climatic feature of an area/ locality
2. Resource availability with the farmers
3. Present level of utilization of resources
4. Economics of proposed integrated farming system
5. Farmers managerial skill
6. Social customs recalling in the locality

Integration of Enterprises

In agriculture, crop husbandry is the main activity. The income obtained from cropping is hardily sufficient to sustain the farm family throughout year. Activities such as dairying, poultry, fish- culture, sericulture, biogas production, edible mushroom cultivation, agro- forestry, agri- horticulture etc. assume critical importance in supplementing their farm income. Assumed regular cash flow is possible when cropping is combined with other enterprises. Judicious combination of enterprises keeping in view of the environment condition of a locality will pay grater dividends. At the same time it will promote effective recycling of residues/ wastes.

Integrated farming system seems to be the answer to the problems of increasing food production. For increasing income and for improving nutrition of the small scale farmers with limited resources without any adverse effect on environment and agro- eco-system. In a cropping system the amount of by products can be as high as or higher than marketable produce. This may go to waste if not utilized in an animal waste component and to reduce the cost of production of the economic produce of component two and finally to entrance the net income of the farm is whole.

Livestock is the best complementary enterprises with cropping, especially during the adverse years. Installation of a biogas plant in crop- livestock system will make use of the wastes, at the same time provides valuable manure and gas for cooking and lighting. In a wetland farm, there are greater avenues for fishery, duck farming and buffalo rearing; utilizing the rice straw and mushroom production can be started. Under irrigated condition (garden lands) inclusion of sericulture, poultry and piggery along with arable crop production is an accepted practice. The poultry component in this system can make use crop the grains produced in the farm as feed. Pigs are the unique components that can be reared with the waste which are unfit for human consumption.

In Rainfed farming, sheep and goat rearing form an integrated part of the landscape, sericulture can be introduced in Rainfed farming provided the climatic condition permit it.

Agro- forestry (silvi- culture and silvi- horticulture) is \the other activities, which can be included under dry land condition. In the integrated system, selection of enterprise should be on the cardinal principal that there should be minimum competition and maximum complementary effect among the enterprises.

1. **Farming System Under Lowland:** common cropping system in rice based system, especially under agro climatic condition of south India is rice- rice – pulse, modified cropping system includes crops like maze, groundnut, sesamum, rice – fish – poultry culture system appears to more remunerative, poultry droppings from the poultry shed placed well above the farm pond meets the needs of fish in the ponded water, Water in the pond can be used for irrigating the crops. About 500 layer chicks are sufficient and excreta can meet the feed requirement of 7, 500 polyculture fingerlings in one ha of pond water rice- fish system is also remunerative.
2. **Upland Irrigated Farming Systems:** Additional income can be generated by enterprises like dairy biogas and silvi culture to the usual cropping systems. Two to three milch cows can be maintained from one ha straw. Recycling of farm and animal wastes through biogas unit can produce cooking gas for family use. Several such integration can greatly increase farm income, besides providing work to family members all through the year.
3. **Rainfed Farming Systems:** Environmentally sustainable dry land farming systems emphasis conservation and utilization of natural resources. Agronomic practices of conservation, tillage and mulch farming, rotational cropping use of legumes and cover crops for improving soil fertility and suppressing weeds and efficient use of cattle manure are some of the components of sustainable farming systems. For above regulation sound land use policy is necessary to tackle the problems of deteriorating natural resources, like soil and water.

Majority farmers maintaining work animals, milch animal, chicks, sheep, goats etc. with crop production. Farmers approaching farming systems, but benefits from these system is low, because of non- educating and also due to non- adoption of improved technology.

Stability in different systems through research

Stability or continuity in production is a crucial characteristic of farming systems whether for the smallholder of the large-scale enterprise. Yet we have not been entirely successful in the past in designing new stable systems. An examination of stability allows identification of its major components and some of these will be discussed in relation to both crop pro auction and animal production systems.

A stable farming system is one in which output shows minimum variation. From year to year and in which this output is achieved without progressive increase in the amount of inputs required. It is probably not necessary to justify the importance of stability. For the smallholder with few resources behind him and often with little access to credit or to cheap loans, a system which ensures a regular food supply, or enough income to buy it, is more important than high average yield. He has little protection against risk. He may well assess and appreciate risk better than the technician who is advising him and the failure of the smallholder to adopt innovations proposed for him can be due to his greater perception of the risks involved. This can be a major and often unrecognised reason for what is described as his "resistance to change". Small farmers are very astute and are not usually slow to take up innovations in which they see real advantage.

Instability: Instability of output is of two types. Firstly, there is the situation where output fluctuates widely from year to year. In many parts of the world, rainfall variation is a major cause, but cyclical changes in pests and diseases and other natural hazards can also be implicated.

Secondly, there is the kind of instability where yield shows a gradual decline over a period of years. In crop production this can, in many areas, be linked to the increasing pressures for continuous cropping systems which population growth has stimulated. Where population growth is a key factor, and perhaps in the tropics generally, the method of farming we most need and still lack is a stable system of continuous annual cropping. Of course there are some excellent tropical examples of stable cropping. Paddy-field rice has been produced on some sites virtually every year for 2000 years or more. Perennial cropping with oil-palm or cocoa, or near perennials like bananas and sugar cane can be equally stable. Shifting cultivation too can be stable, though in terms of land-use it is hardly continuous.

The stability of these systems is dependent on their ability to meet two essential needs. First is the avoidance of erosion and secondly the ability to provide a continuing nutrient supply adequate for the level of production, either with or without the use of fertilizers. In paddy rice production, the landscaping and the standing water, and the supply of biologically fixed nitrogen are the major factors creating stability. Perennial crops mimic the natural vegetation in providing continuous cover of the land, so restricting erosion. Their high transpiration rates reduce leaching of nitrogen which in an, case is only briefly in mineral and therefore leachable form as it cycles within the crop: veil system. But where annual rain-fed cropping is the tradition, shortening of the resting phase in shifting cultivation can be disastrous because of the yield decline which may follow. Nor is this just a problem for the small farmer. There have been several attempts at large-scale continuous mechanised maize and grain legume production in Ghana over the years. Most if not all have been discontinued because of declining yields. Just as the small-holder without resources is dependent on a stable low-risk system, so the large scale mechanised producer is dependent on maintaining yields at a level which will meet his high input costs and still leave some margin. Declining yield due to steady erosion robs him too of stability, and it cannot be reversed by fertilizer use alone.

Crop residue mulching and zero tillage: This is why innovations which are under study at IITA in Nigeria are of such great interest to both types of producer. The technique is based on mulching with residues from the previous crop, coupled with zero-tillage planting techniques. Mulching is no new practice and its benefits in terms of water conservation, erosion control and soil temperature modification are well-known. Previous practices have used grass and other materials brought to the field, but this requires a great deal of extra work and the availability of an area from which mulch can be obtained.

Planting through the mulch without cultivation can be done by the smallholder using a modified "Jab-planter". Zero tillage drills are available for the mechanized farmer. The mulched and uncultivated soil has a greater organic matter content and more stable structure, with adequate permeability provided by the undisturbed root holes of previous crops and the increased population of soil fauna. Erosion is controlled, nutrient losses are reduced and the nutrient holding capacity of the soil is increased.

Desertification: In livestock production, instability of the type I have described is perhaps most obvious in the drier parts of the world. One example is the increasing desertification occurring on the fringes of the Sahara desert. Desertification can be defined as the development of a landscape of shifting sands and little vegetation in areas which did not previously have these characteristics. The southern border of the Sahara has in many areas moved 100 km south between 1958 and 1975, due to the effect of increasing human population on the ecological balance. Inappropriate methods of crop production and the removal of woody vegetation for fuel are contributory causes but overgrazing is the main factor resulting in the agricultural output of some areas falling virtually to nothing. In these communities, welsh is often measured in terms of herd size rather than its quality or output and this is an attitude which will be slow to change. It is difficult to avoid the 'numbers' approach and to achieve appropriate grazing management where land or grazing rights are in common ownership, So schemes to develop individual or small-group rights aver land or the leasing of land, for example by Botswana in another dry part of Africa, are a step in the right direction. But even without these extreme conditions, rainfall fluctuations are a major cause of instability or of wasteful feed utilisation. The farmer cannot easily adjust his livestock numbers and so his stocking rate will usually be kept at a level to match the feed supply in years of shortage, Even in England, it has been estimated that one third of the grassland in the south-east is essentially a reserve against drought. In all but the two or three driest years in ten it is under-utilized or not needed. So if livestock are to be kept at reasonable intensity and stability, the feed supply must be buffered against rainfall variation; by using irrigation, by having a reserve of stored feed, or by using crops which show some drought resistance. It is likely that a system based on cane production will be more tolerant of drought than one based on pasture grasses which have to be cut or grazed at frequent intervals and which for various reasons are therefore more prone to drought effects.

Nitrogen inputs: In addition to reducing yield fluctuations from year to year, and decline in yield over years, one should also try to buffer farming systems against economic instability. This can partly be achieved by diversification and the effective integration of different, perhaps crop and livestock, enterprises. But it also means making predictions about likely future changes in market demands and prices, and in the supply and cost of inputs. The input factor of particular and world-wide interest today is the likely future trend in the cost of nitrogen fertilizer. Industrial fixation of nitrogen is very costly in energy terms so that as energy supplies become scarce and more expensive, perhaps from the mid 1980's, so fertilizer nitrogen will rise in cost. Indeed, this trend has already started. In intensive farming, the support energy used to make nitrogen fertilizer can be as much as 30-50 per cent of the total support energy used. Of course, farming is a minor user of support energy compared with cooking transport and other industries but this will not prevent nitrogen fertilizer becoming more costly. The off take of nitrogen in livestock production is not high. An intensive enterprise producing 8000 litres/ha of milk is removing from the farm only about 40 kg/ha of nitrogen. But if fertilizer is the only source of nitrogen for forage production, the farmer may have to use as much as 400 kg/ha of nitrogen as fertilizer, or ten times the amount he removes. So, future stability in both farming and food prices will in part depend on economies in the use of nitrogen. This has four implications.

First, we must seek to make more effective use of fertilizers; we commonly recover in the crop only about half the nitrogen we apply, at best about two-thirds. This means better timing and more accurate prediction of the optimum application. Secondly, it may be possible to reduce leaching and denitrification losses. Thirdly, we must make more effective use of farm wastes. Lastly, we need to make greater use of biological fixation especially that provided by legumes and, for livestock feeding, particularly the forage legumes.

Forage legumes: Twenty five years ago forage legumes did not appear to have much of a place in tropical livestock production. One could hardly take that view now, even without the threat of increasing nitrogen cost. The first point to make is the wide range of morphological and ecological types now available. They show some variation in response to temperature and to water supply, though none are adapted to conditions of poor drainage. There are woody perennials, herbaceous perennials, and self regenerating annuals. There are procumbent and stoloniferous types tolerant of heavy grazing, and trailing and twining types which are better adapted to competing with tall-growing grasses. Most of the forage legumes are not very tolerant of shade, but there are a few which show some adaptation. Many of the commercial species are self-fertile and therefore show good varietal stability. A few like *Desmodium heterophyllum* are good grazing plants, but have poor seed production characteristics and have to be vegetative propagated. Apart from some 'hard seed' problems most of the forage legumes can readily be established from seed, though early growth is usually slower than the grasses.

It must therefore be an important task in developing ruminant production systems to examine the suitability of different forage legumes for the local circumstances of:

a) Soils and climate

b) Different utilisation needs, e.g. grazing, cut feed, conservation

c) For integration with other local feeds, both in terms of seasonality and nutritive value.

The forage legumes often show a lower acceptability to livestock than the grasses, but even this can sometimes be an advantage. Prime need in beef production was an improvement in the quality of dry season grazing, He achieved this first through the foliage growth of lopped browse trees but he also studied a number of forage legumes which might be sown into existing natural grazing. *Stylosanthes guanensis* proved the most appropriate: it was more productive than the others he examined, it was partly rejected in favour of the grass component during the rains and so its foliage remained to improve the quality of dry season grazing. Thus a situation in which indigenous cattle on natural grazing lost weight for five months of the year and gave an output of 100 kg/ha/yr was altered on legume improved swards to a weight loss for only three months of the year and an output of 180 kg/ha/yr. Another development of importance here is in our ability to sow improved species into existing pasture. Frequently, the better classes of land are best used for cash and food crop production and much of the live stock feed must come from natural grazing on land which for one reason or another is difficult to cultivate. Equipment is being developed in different parts of the world which can achieve this, using various sod-seeding techniques with or without local application of an herbicide, according to the conditions.

Nitrogen fixation: If we are to get the maximum out of forage legumes we must pay special attention to the process of nitrogen fixation. Temper ate species seem in general to be well-supplied with reasonably effective naturally-occurring strains of Rhizobium. It is very common to find field-grown forage legumes in the tropics infected with strains of low effectiveness. So any legume programme must be accompanied by nodulation studies. The variation in effectiveness can be very large. The dry matter yield of the clover varied with the isolate by as much as a factor of eight, and the mean yield of all of them was only half that of plants inoculated with a commercial strain of Rhizobium. Thus, inoculation may be a very important part of pasture establish meet. The entry of Rhizobium organisms is an infection process, and once infection by one strain has occurred the legume may be virtually immune to others. Unfortunately, the most aggressive and competitive strains of Rhizobium are not always the most efficient nitrogen fixers. Where a naturally-occurring strain is aggressive in infection but poor in fixation, inoculation with a better strain can be difficult. New methods of inoculation may increase the competitive ability of the inoculum strain used. There are many examples in the literature of fixation rates in forage legumes of 100 to 200 kg/ha/yr of nitrogen. These and greater quantities should be possible if we give attention to the matters discussed above, and to the whole question of selecting compatible combinations of legume cultivar and Rhizobium strain because this is an area which has received so little study in the past.

Innovations on the farm: The methods of farming we develop must not only be in tune with the local conditions of land and climate and the local economic conditions, but they must meet the aims and circumstances of the farmer, must avoid unacceptable risks, yet exploit his entrepreneurial abilities. In transferring technology to the farmer, it is particularly important to know whether the response to an innovation or whether the output from an enterprise is likely to be the same on the farm as it is on the research station.

Approaches within farming systems research

The term FSD often is used loosely. An activity is considered to be a legitimate part of the technology research function of FSD if the following situations hold:

- The whole farm is viewed as a system.
- Research is conducted with a recognition and emphasis on the choice of priorities that reflect the whole farm.
- Research on a farm sub-system is legitimate FSD, provided the connections with other sub-systems are recognized and taken into account.
- Evaluation of research results explicitly takes into account linkages between sub-systems.

As long as the concept of the whole farm and its environment is preserved, not all the factors determining the farming system need to be considered as variables — some may be treated as parameters or constants.

A distinctive feature of FSR, or rather FSD, is that a systems perspective is constantly borne in mind. Within this framework, one can look at either a small or a large number of variables in the system. In general, because of the complexities of simultaneously handling a large number of variables, most FSR programmes have tended to limit the number of variables they study and to regard the other factors that influence the farming system as parameters or constants.

Progression in farming systems thinking

Three general approaches to FSR or FSD with a technology generation focus, have been defined relating to the number of variables being investigated. These approaches can be described as FSD '***with a pre-determined focus***', '***in the small***', and '***in the large***' In a sense these types represent an evolution in FSD as techniques have been developed to handle progressively more complex situations — defined as involving a higher ratio of variables to parameters, The progression has been expanded to four phases, can be articulated as follows:

- In its early days, FSR (i.e., FSD) type activities concentrated on how farm yields of particular crops could be raised. This form of on-farm research stood in sharp contrast to earlier multi-locational on-farm trials, in which only the effects of the physical (i.e., natural or technical) environment were tested, Even though this FSD approach did involve the inclusion of socio-economic elements and, hence, had a farming systems perspective, it was done with a 'predetermined focus' on the productivity of a particular commodity. Thus, this approach involves looking at one facet of an enterprise or one specific enterprise and identifying improvements within that focus that are compatible with the whole farming system. With the creation of distinct FSD teams — in contrast to an on-farm testing component located within each station based commodity team — FSD activities generally have evolved beyond this level. However, it is important to recognise that this type of activity still is a legitimate part of any FSD in the sense that it can provide a very useful input into the activities of commodity research teams.

- In the pursuit of greater farmer participation, research shifted to addressing farmers' articulated problems directly. Still, farmers' problems were very much con fined to crop production, although links with other components of the whole farm system were studied. An alternative name for this 'farming systems in the small' focus could be '***farming systems with a whole farm problem focus***', Many FSD efforts currently tend to be at this point, However, currently, on-farm research projects that seek to improve whole farm performance through manipulation of the linkages between all enterprises are few,
- Looking further ahead and to enable addressing issues of sustainable agriculture effectively, concerns for maximizing the economic and biological performance of enterprises will be modified by concerns relating to ensuring sustainability, including the rehabilitation and regeneration of natural resource systems. This natural resource system dimension to sustainable agriculture can be thought of as a type of 'farming systems in the large, or '***farming systems with a natural resource systems focus***',
- Similarly, attention to ecological sustainability will need expanding to include the many off-farm and non-farm activities that make up sustainable livelihoods. This also can be thought of as a type of 'farming systems in the large, or '***farming systems with a livelihood systems focus***'.

These notions, particularly the last two, have been largely unexplored until recently, when techniques have evolved to handle complex situations in which the ratio of variables to parameters is high. Formal modelling techniques by themselves generally have limited value in addressing such situations because of the complexity of the relationships and the degree of understanding that is required initially to develop a realistic model. However, the major breakthrough that has occurred is the ability to pursue a less formal or more informal modelling approach through application of farmer participatory type techniques that, in essence, involve using farmers' minds as computers!

FSD with a 'pre-determined focus' on specific technology

Researchers in station-based programmes often want FSD workers to evaluate specific technologies they have developed to see how they fit in with constraints of the farming system currently used by farmers. This type of work is referred to as FSD with a 'pre-determined focus' or 'predetermined technology focus,' In Botswana, FSD teams and participating farmers have assisted station programmes in testing cowpea varieties, evaluating the desirability of hilling groundnuts, ascertaining whether the Dutch hoe can help with the weeding operation, and so forth.

This type of testing in FSD is not analogous to typical technical transfer. Rather than testing fully finished products, FSD assists technology generating programmes, usually based on research stations. Often, the types of technology tested in this manner have not been identified as having high leverage power in the system. Leverage power refers to the ability of a technology change to have a major impact on the performance of the farming system. However, these technologies do fit possible needs and asking farmers to participate in their development is worthwhile, It is desirable to test technologies as early as possible in their development, This is to facilitate making modifications and to minimize waste of research resources. It also may accelerate the dissemination of suitable technology, such technology, even if not fully appropriate, may sometimes stimulate thinking among farmers if they are not accustomed to these types of interventions. Feedback in these instances can contribute to new innovations in technology generation.

FSD with a 'pre-determined focus' on a subject

A great deal of FSD work, particularly in its earlier years, was not focused on testing specific technology as much as a pre-determined focus on a subject matter area. FSD within the CGIAR institutions such as IRRI, CIMMYT, and ICRISAT have focused on issues related to production of their respective commodity crops. ILCA focused on issues in livestock production, ICRAF in agro-forestry' etc. Some FSD teams within the national agricultural research systems (NARS) also worked under limited mandates to deal with particular enterprises; commodities; or other subject areas (e.g., soil water management). The judgements of farmers, the end-users of products developed by these organizations, were foremost in these FSD activities. However, the parent institution with its technology generation activities would be a primary client in these instances.

FSD with a 'whole-farm problem perspective, or 'FSD in the small'

An FSD interdisciplinary team conducted a diagnostic survey and determined that labour for satisfactory weed control in field crops is a major constraint to farm production and farm family well-being. The survey also showed that due to population pressures, farmers were no longer able to shift field sites as in the past and that deteriorating soil fertility also was becoming a major issue. However, during the course of farmers participating in testing possible solutions to these problems, discussions between the FSD team and farmers identified the potential for intensified poultry production as an additional means of improving farm income. This resulted, in addition, in collaborative testing in the area of intensive poultry production and, as a result, exemplifies FSD with a 'whole-farm problem perspective.'

The major breakthrough that has occurred is through supplementing the rapid rural appraisal (RRA) techniques, which have been available for some time. To date, the first two approaches listed above have been used. Although FSD in the small arrives at a focus within the system in the course of diagnosis, FSD with pre determined focus moves into the system to research an enterprise or one facet of an enterprise looking for improvements within that focus that are compatible with the whole farming system.

FSD with a natural-resource management focus

A non-governmental organization (NGO) has adopted, as one of its aims, the conservation of the natural resource base within areas of agricultural production. This NGO sponsors FSD activities in a village in which participating farmers look at current and prospective problems in natural resource conservation and management. They evaluate potential solutions to these problems at the same time issues in agricultural production and family income are considered. The FSD team from the NGO and farmer participants discovered a number of in congruencies between issues of conservation and present day agricultural production. They included the following:

- Natural-resource conservation analyses focus on the landscape (e.g., possibly a watershed), which in some instances does not even correspond to a village or to any other political entity. On the other hand, agricultural production analyses tend to focus on individual farms, units within farms, or smaller interacting networks of farms.
- Conservation invariably takes a long-term perspective of problems and solutions, whereas production analyses consider short-term returns and immediate survival issues.
- Natural resource conservation often involves community or ultra-community efforts and goals, which might conflict in terms of labour or other requirements, with individual farm production activities.

Integrating these apparently opposing sets of objectives for natural resource conservation and production poses a major challenge for FSD. Participating farmers recognize many of the conservation issues and trends and some express concern for the future. In some cases, traditional techniques for conservation such as gully amelioration have been identified and are being promoted, but other issues remain unresolved. Some of the constraints encountered today or anticipated in the future were unknown to farmers in the past. Solutions therefore, are not known, and innovative and problem-solving thinking is required. However, given farmers' intimate knowledge of their current and past production environment, their input in designing appropriate strategies for the future based on an analysis of what has happened in the past is indispensable. This is clearly a fertile area for active collaboration between on-farm and on-station work: a give and take between farmers, scientists, and policy makers.

Some characteristics of FSD

Some of the major characteristics of FSD are as follows:

- **Farmer Centre Stage**. The farmer, as the consumer of the improved technologies, is in the centre of the stage. This provides an opportunity for researchers to learn from the farmer, enables him or her to have an input into the research process, and ensures that criteria relevant to him or her are used in evaluating proposed technologies. For the farming family, evaluation criteria (i.e., for the adoption of the improved technologies) can be divided into the following groups:
- Necessary conditions determine whether the farming family would be able to adopt the improved practices. Such conditions include technical feasibility, social acceptability, and compatibility with external institutions — that is, support systems.
- Sufficient conditions determine whether the farmer would be willing to adopt the improved practices. Obviously, the necessary conditions will be influential in determining this willingness. Sufficient conditions include the compatibility of the improved practices with the goal(s) (e.g., food self-sufficiency, profit maximization, risk minimization, etc.) of the farming family; the resources they have access to; and the farming system they now practice.
- **Work with Representative Farmers**. Although the input of farmers is critically important in FSD it is impossible for FSD teams to work with all farmers. Therefore, a few are selected that are thought to be representative of all farmers. Because there are many types of farmers — with differences in the products they produce, the resources they possess, and the problems they face. It is necessary to put farmers with similar characteristics in the same group (i.e., stratum or recommendation domain). A small sample of farmers in each of these groups then is selected to work with the FSD team. If the grouping and selection of farmers are carried out correctly, then the results obtained by these representative farmers should be achievable by other farmers with similar characteristics, when the technologies are extended to other farmers by extension staff.

- **Involves an Interdisciplinary Approach**. As indicated earlier, farmers have complex farming systems. As a result, changes in one part of the farming system may have a good or bad impact on another part of the farming system. For example, in some areas, it is not realistic, or indeed desirable, to try and increase crop productivity without considering livestock at all,' Also, attaining a successful solution to a technical problem that has been identified will depend on whether the farming family has the labour and money to adopt (i.e., an economic issue) and any sociological reasons preventing them from adopting. Therefore, to address the wide range of farming systems' issues, FSD teams generally consist of representatives of a number of disciplines — usually agronomists, animal scientists, and agricultural economists and sometimes sociologists. As a result, an interdisciplinary approach, that is a number of disciplines working together on the same problem, has to be used to solve the problems of farmers.

Dynamic and Interactive Appraoch. Because of the dynamic nature of agriculture, research is a never-ending process. Farmers always face problems to varying degrees in their farming operation, and many of these problems can be solved through research. Sometimes the solutions suggested as a result of research don't work and need to be modified. This is implied in Figure 3.1, where the dotted lines indicate that such failures require moving back to an earlier stage in the FSD process and repeating one or more of the steps in the process. This makes FSD an iterative as well as a dynamic approach.

Complementary to Station-based Research. In the context of technology choice and development, the role of FSD (i.e., utilising a systems approach) is seen to be complementary to technical component research (i.e., which primarily employs a reductionist approach), most of which is undertaken on experiment stations and is usually commodity-based, With respect to such research, FSD has three roles:

- To look at recognized farming systems and the stock of materials and techniques accumulated from station-based research, so as to be able to choose technical solutions to problems that have been identified, On-farm experimentation then adapts chosen solutions to the local situation. This is a mobilizing and adaptive role, shaping the product for an identified market.
- To pass back unsolved technical problems, important to the system, to the appropriate commodity research team on the experiment station, This role is one of identifying and helping to prioritize the agenda for technical research.
- To link with farmer clients and extension staff in local farm situations, drawing both farmers and extension workers and other relevant 'actors' into the technology-generation process.

FSD with a 'livelihood-systems focus'

There are, to date, relatively few good examples of applying FSD with a 'livelihood systems focus.' One organization that in recent years has expended considerable effort in integrating 'off farm' with 'on-farm' activities in a sustainable manner has been ICLARM which has concentrated on integrating aquaculture with crops and livestock.

Certainly integration of 'on-farm' and 'non-farm' activities deserve much greater emphasis than it has to date, especially as increasing pressures develop on the natural resource base. 'Non-farm' activities car provide an outlet for surplus labour and contribute to the attainment of sustainable livelihoods. In many areas (e.g., semi-arid parts of West Africa), particularly where agricultural activities are seasonal in nature, farmers have always recognized the complementarily between farm and non-farm activities, through emphasizing the latter particularly during slack periods in the agricultural cycle. Many of these jobs are in the non-formal sector and involve micro-enterprises also an area that needs much greater emphasis than it has to date.

Technology that works under experimental conditions will not always be relevant on-farm

In Botswana, precipitation is low and variable, creating erratic conditions for crop production. Under these conditions, farmers minimize risk by adopting an extensive low-input cropping system that is moderately productive under favourable conditions but minimizes losses when rainfall fails. Under research-station management, an intensive system for maize increased average yields and dramatically improved yield stability. The intensive system consisted of controlled pathways for planting with minimal tillage of the soil and water concentration onto the plant zone. It is the water concentration that stabilizes maize yields. Weed control was as used needed but weed pressures were minimal when crops were established early.

This technology failed when it was tested under farmer management. It failed because critical resources are not available on the farm at the right time to meet the schedule requirements of the new technology. The intensive system requires that soil preparation be finished before the rainy season begins and that planting be carried out on the first rain(s) of the year. Meanwhile, draught animals to do this work are kept near watering points at distant cattle posts until rains have filled shallow wells and ponds near the herds. Farmers and their draught animals cannot plant early in the season, and when they do not, weed growth overruns the controlled pathways. The Ministry of Agriculture made a fitting decision to conduct farmer-managed tests before recommending this system to farmers based on station results alone.

FSD in the context of agricultural institutions

Work relating to FSD might be incorporated as a working tool for several different types of institutions. These could include such diverse institutions as agricultural research, extension, community development (particularly in regards to NGOs), or even ministerial or national units interested in policy design and assessment.

FSD thus can he used to focus on family income, technology design, dissemination, and adoption: or designing solutions to the problems associated with a deteriorating natural resource base in agricultural areas. At another level of viewing the institutional setting of FSD, it is apparent that not only some disciplines but also various technical and sociological approaches and methods, are evolving to better deal with systems analysis. Take, for examples, the use of Geographic Information Systems (GIS) as a tool integrating large amounts of data into a single framework useful for technical as well as sociological and economic analyses. On the other hand, community development and policy reform programmes use increasingly sophisticated techniques (i.e., tools or methods) that contribute to decentralized planning, grass-roots participation, and empowerment in assessing problems and designing solutions.

The emphasis in many areas is on developing integrated approaches and system type analytical methods. With the possibility of increasing convergence occurring in some of the methods and, in some cases, objectives, the unique contributions by FSD will be brought into question by some. And, in fact, the institutional role or mandate of FSD will vary. FSD in one setting may focus largely on trials and farmer testing of new technology. In another setting, a greater emphasis may be placed on surveys, sociological or information related to the financial picture of the whole farm, and so forth, in the larger picture, none of this matter too much. But it should be remembered that FSD does have certain fundamental characteristics.

These include:

- Iteration, such that no single proposal should be considered definitive.
- Integration, usually involving multiple disciplines and explicit building on managerial perspectives and managerial values.
- A consistent orientation towards problem solving and an attitude that no perspective on this problem should be discounted.

FSD also must focus on appropriate linkages with other development support programs and institutions. In this regard, FSD emphasizes a strongly integrative and integrating philosophy, not just for disciplinary and farming perspectives within FSD itself; hut integrating strengths of personnel, resources, information, and location of the FSD programme with other research and development programmes working in the same area or domain. A critical element of this inter-institutional integration is a logical division of labour for the problem-solving task; this invariably will encourage collaboration in the development of work plans both for short-term as well as medium- and long-term objectives. A number of challenges face FSD if it is to continue playing a significant role in facilitating the process of agricultural development. Some of these are:

- **Better Incorporation of Farmers**, Incorporating farmers into the research process was one of the most important principles underlying the evolution of the farming systems approach. The basic justification for this was that farmers could improve the efficiency of the research process. At a minimum, it was argued they could prevent the use of research resources on types of technologies that they would be unlikely to adopt, while even more importantly, it was anticipated they could, in their own right, contribute positively to the research process. Unfortunately, farming systems workers often have not sufficiently recognized this positive and interactive contribution of farmers. Consequently, this spawned the farmer participatory research (FPR) movement. The problem of not utilizing the farmer sufficiently in FSD activities is not due to inherent deficiencies in FSD itself but rather with the way in which it has been applied, Techniques of FPR or PRA need to be incorporated into FSD.
- **Continued Evolution of FSD**. FSD is relatively new. Therefore, the methodology is still evolving and, as a result, universally accepted 'standard texts' on the 'nuts and bolts, of how to do it are still to emerge. Related to the methodology issue is that time and cost-efficient (i.e., money and people) methods for undertaking FSD still need further development. This is important because of the limited resources available for undertaking research, both on-station and on-farm.
- **Greater Incorporation of the Policy/Support System Perspective**. As has been stressed already' the farming systems approaches implemented to date have focused mainly on the technology dimension (i.e., FSR). However, a basic principle of the FSD approach is that the farming systems perspective is critically important in formulating and adapting policy/support systems in ways that will facilitate and accelerate the agricultural development process, Once again, discussions relating to the policy/support system are scattered throughout the manual and it is highlighted as an essential component of FSD. Failure to incorporate this dimension in FSD activities is likely to have an impact analogous to playing soccer on one leg, Nevertheless, because of the lack of much proven experience and documented material on the farming system perspective with respect to the policy/support system, this manual does not deal with this dimension to the extent that would be desirable. Hopefully, this lack of experience will be rectified in the fairly near future.

- **Incorporating Equity Issues — Intra and Inter-Generational**. FSD tries to help the farmer with the problems he or she has identified. Of course, the reason for this is the necessity to introduce an intervention in which they are interested, It's likely that these felt, problems of farmers are likely to have a short-run focus (i.e., particularly when the farmer is operating very close to the survival level). Also, it is possible that helping to solve the individual farmers' problems creates others for the society as a whole. It pertains to inter-household relationships, equity considerations within a particular generation also apply to what is happening within farming families. As will be discussed later, it is wrong to assume that all farming households or families operate in such a way that distribution of effort and benefits is equitable. When designing a technology to help farmers increase their productivity, consideration must be given to the possible long-term effects (e.g., decreasing the amount of productive land available) of that intervention. Therefore, if farming system (FS) workers are not careful, their work can result in creating two types of inequalities, that is, helping some farming households — or even certain individuals within those households — at the expense of others and/or reducing the quantity and quality of land that can be productively farmed by future generations. Avoiding the development of such inequities constitutes a major challenge to FSD teams who need techniques to be able to screen out proposed technologies and policy/support systems that would encourage or exacerbate such trends. Once again, issues relating to equity matters, although generally recognized as important, still need attention in farming systems type activities. This will require continued evolution of FSD towards the FS 'in the large' approach.
- **Assessing Agricultural Research Impact**. Related to the research resource issue is the importance of devoting some effort to assessing the impact of the research process - something that often has been done inadequately. More attention needs to be paid to adoption/diffusion studies. Such studies, of course, are the best measure of the impact of the agricultural development process, Therefore, the credit for favourable results from such studies cannot be allocated to FSD activities alone. However, FSD teams, because of their on-farm location and pivotal linkages with other actors, are often in the best position to take a leadership role in their execution. Also, as suggested earlier, such studies can and should be used for other purposes, such as feeding back priorities for further research and providing evidence for adjustments in the policy/support system to encourage greater adoption,
- **Improving Credibility of FSD**. Establishing credibility for FSD-related activities is a major challenge and is necessary to ensure that some of the limited research resources always will be allocated to them, FSD staff

constitute only onc set of the 'actors' in the agricultural development process. Additionally, FSD helps facilitate a process and does not result in a product by itself. Thus, FSD cannot claim sole credit for any technologies developed for, or adopted by, farmers. However, it achieves credibility through its linkages and cooperative efforts with other 'actors' in the agricultural development process. These 'intermediate products' need to be documented and publicized to a greater extent than often has been the case.

13

Crop Diversification for Sustainable Farming System

Agriculture has under gone drastic changes from the mid 1960s through the introduction and application of various newly developed techniques in agro science such as crop and water management practices, Emphasis has been given to be integrated system approach in crop production.

It is essential to plan and implement rational policies to minimize the regional imbalances of agricultural production. The present growth in crop production has to be sustained considering ecological and economic factors. The existing as well as emerging frontiers must be examined to enhance agricultural growth on a sustainable basis. The understanding of agroclimatic relationship through crop weather conditions, soil fertility, water use efficiency, rain water management and appropriate cropping patterns based on regional resources potential must be emphasized more over agro forestry, integrated crop management, biotechnology and use of renewable energy must be emphasized for use in environmentally harmonious agriculture. Training of farmers has to be adapted to be changing technological environment.

The forest affects soil and water in many ways. Leaves and branches decreases wind strength, alter the absorption of solar radiation, and increase the surface area for evaporation. The root growth and the decomposition of plant organic matter by soil fauna and microbes, modify soil texture and structure, affecting water penetration and drainage. Changes in vegetation cover can therefore have a significance influence on the hydrological cycle and climate system.

The study area consists of the state of Punjab, Haryana, Himachal Pradesh and Jammu & Kashmir. On the basis of physiographic and soil resources, this zone is clearly divisible into two sub zones; the plains of Punjab and Haryana and the highlands. Based on altitude soil type, geology, and rainfall each of these two sub zones may be further subdivided. Major sources of water in the region, in addition to rainfall and river, are lakes and ice crops of glaciers and undergrounds springs. Under ground water is largely concentrated in the plains of Punjab and Haryana.

The region covers a varied land use, from highly intensive monocultures through rotational and intercropping to grazed rangeland and near natural forests.

Land Use Change

Our analysis of land use for the last 40 years in the four states suggests that Punjab and Haryana have reached the absolute limit of expansion of area under cultivation with almost 84% of the area being cultivated. Six of 8% of the area of these states is under urban uses. Another 5% are under forests (mainly strip forests) and the remaining 2-3% is roads, canals and other infrastructural and industrial uses. Cultivable waste as a category has virtually disappeared in these two states. Such intensive land use for agriculture is sustainable only with increasing and continued high doses of balanced nutrients and other inputs such as chemical fertilizers and insecticides. The proportion of area available for cultivation in Himachal Pradesh and Jammu & Kashmir because of topography and physiography is rather small and cannot be expanded without major private and public investment that in return will result in major ecological problems and should be avoided.

The cropping pattern in the region has under gone a substantial change, with wheat and rice emerging as a major crop rotation in Punjab and half of Haryana. Its expansion in Himachal Pradesh and Jammu & Kashmir has been moderate. Crops that have been replaced by wheat and rice are gram, bajra, barley, millets, and pulses. Area under cotton has grown in Haryana. In absence of expansion of the sugar industry, the area under sugarcane has remained static. The cropping pattern of the region has unnecessarily become energy intensive and is affective the static balance of the underground water resources in the plains of Punjab and Haryana. The growth of infrastructure irrigation and other technological factors are responsible for a major shift in cropping pattern in favour of wheat and rice in the states of Punjab and Haryana.

A number of policy steps must be taken to encourage farmers to switch from rice, which is a water and fertilizer intensive crop in the region, to crops that demand less water. This can be achieved through price policy research and development efforts, and establishment of agro processing industries as so to make sustainable alternatives more attractive to the individual farmers. Without supplementary organic manure, intensive agriculture leads to depletion of soil fertility. Ludhiana district in the green revolution state of Punjab, which records the highest yields of many crops, now also records the highest deficiencies of plants micronutrients. Extensive use of organic manure is the only way to overcome the deficiency in Punjab, above 5 million tons of rice straw is being burnt every year during October to December, if crop residues were ploughed back into the soil, the rate of micronutrient depletion would be substantially reduced.

Since 1965 , when water from the Bhakhra canal was brought to the farm, the rise of the water table has also been a serious phenomenon since 1985, the rate of rise in the water table has been above I m annually Patches of salinity have started appearing at the farm level. The situation is worse in higher rainfall areas where water logging follows shortly after the rains. Apart from affecting agricultural crops, a high water table causes floods even during slight rains because of the reduced moisture capacity of the soil, In Hisar, the bearing strength of the soil has declined to less than 50% in 50 years.

Diversification of Agriculture

A diversification of agriculture to increase the area under oilseeds and pulses should be encouraged. Sunflower is becoming a prominent crop among the oilseeds. Its water requirement is quite high. Although the sugarcane area has been substantially increased, it has not reduced pressure on ground water. The ground water position has been distributed by a tremendous increase in food production, especially wheat and rice (119.2 lakh tonnes in 1980-81 to 192.14 lakh tonnes in 1992-93).

The ravines and undulating areas of Kandi tract can also be developed for horticulture which will reduce spending on reclamation. The development of land first for agriculture and then its conversion for horticulture is not being appropriate method. Horticulture requires less water than an intensive cropping system.

More area is brought under cotton because its return per hectare can compete with those from paddy. Crop rotation which is a fertilizers process should be promoted in a holistic manner. About 7 to 8% area should be cultivated under fruit trees and marketing facilities provided.

Potential for Diversification of Agriculture

The economics of Jammu & Kashmir and Himachal Pradesh have large forestry and horticulture sub sectors. Forest area and forestry development area in both these states are substantial. The growth of different types of forests during the last two decades has been uneven; the area under forests in Punjab and Haryana is less than 5% but is slowly growing. There is not much scope for growth of block forests in these states. Most of the growth has been in strip forests on the banks of canals.

Despite major data problems for a temporal analysis of forests cover in the study area, we found that as per official records, 33% geographical area of Himachal Pradesh, 9.9% of Jammu & Kashmir, 5.6% of Punjab, and 3.8% of Haryana were under forest cover in 1986-87. On per capita basis the lowest forest cover is in Haryana and the highest in Himachal Pradesh, in absolute terms total tree cover is 20,880 sq. km in Jammu & Kashmir, 12882 sq. km in Himachal Pradesh, 776 sq. km in Punjab, and 644 sq. km in Haryana. The regeneration of forest for Himachal Pradesh and Jammu & Kashmir has been observed and needs to be monitored more carefully. A study of various forest types in combination with horticultural, pastoral and other systems suggests that there is a wide variation in expected returns per year per hectare, which seems to be more attractive than those from crop husbandry provided marketing is taken care of.

Establishment of agro forest processing is required on a regional basis rather than on a state basis. The problems of marketing forest products, particularly wood in the absence of such industries with in the region discourages individual producers from under taking this activity.

Integration of Horticulture with Agriculture

Area under horticulture in all the four states has been growing rapidly since 1970-71 with fastest growth in Himachal Pradesh. The economics of Himachal Pradesh and Jammu & Kashmir have a significance horticulture sub sector that is growing and emerging as a major component of the agricultural and agro processing facilities are some of the important problems of this sector.

Himachal Pradesh and Jammu & Kashmir are considered the fruit baskets of the region because of favorable climate and topography. In Punjab and Haryana, area under horticulture is small (less than 1%) but increasing. Some illustrative measures of rates of returns from horticulture suggest returns ranging from 30 to 40 %. However, the experience of marketing particularly of apple, suggests that expanding wood demand for packing of fruits is creating serious stress on the forests, especially silver fir and spruce. This has already been noted by the government of Himachal Pradesh and Jammu & Kashmir Subsidized card board boxes are being experimented with.

In view of the climate and the rugged terrain, fruit production is the only highly profitable enterprise where crop growing is not of much utility. Moreover, horticulture enhances the cohesiveness of the soil, preventing soil erosion.

With increase in demand for fruits, the National Commission on Agriculture has indicated that production must increase. Farmers thought, therefore, to be induced to grow more fruit trees. Induction of farmers is possible, once a complete understanding of soil economics is arrived at.

The following possibilities are favorable to horticultural activities in the zone:

i) Agroclimate and topographic conditions favour horticulture as an excellent source of income per unit of land area.

ii) Horticulture helps in using the land more efficiently than crops and conserving the soil, which is highly susceptible to erosion in case of cultivation.

iii) Horticulture permits the maximum use of natural resources by adopting the negative propagation.

The actual impact of horticulture on environment is increased by the need for packing cases. The use of wood for packing fruits has greatly increased the burden on the forest wealth, leading to extensive deforestation.

The standard boxes for packing apple can contain 9 to 18 kg. However, 1 ton of the fruit is supposed to be contained in 55 standard boxes. Apart from the wood required to manufacture the boxes, 25% is further wasted on manufacture of logs and billets, 10% on sawdust, 15% on cut off rejection. Thus, one -third of the standing volume is wasted. About 65 packing cases of standard size are obtained from 1 cu m of the silver fir and spruce forest. The trend indicated that annually 10.8 km^2 of forest is lost while 6 km^2 is raised; a tree must grow about 100 to 120 years before it can exploited. The result is deforestation. Deforestation in turn leads to land degradation and soil erosion. To curb the problems and have an economically feasible and environmentally sizes of packing cases, use other materials for packing, and use wood in other forms.

The Himalaya is now under tremendous transformation that is accelerated with enormous speed. Most of the original lush green natural vegetation is presently replaced by shrubs, savannas; grass accompanied by gullies, ravines, and eroded and scraped landscape. The major use of forests presently is to meet the demand for industrial wood, which is needed in a great quantity, a large part of which is used for packing horticultural products.

Native plant cover normally, provides good erosion control, hence sustenance of natural resource base for economic growth and development compatible with environment, problems arise, though, when an area is deforested and converted into cropland, increasing soil erosion. The resulting soil loss reduces soil productivity and modifies the environment.

Cropping systems in Haryana for sustainable resources

1. Paddy - potato - onion- vegetable
2. Paddy - berseem- sorghum- wheat/ field pea
3. Fodder - toria -wheat / onion
4. Sorghum - (fodder) - berseem
5. Paddy - wheat
6. Paddy - wheat - GM/ Green fodder
7. Paddy - raya- s'cane - ratoon

8. Paddy - potato - toria- sunflower
9. Maize- potato - onion/ vegetable / sunflower
10. Sorghum/ maize (fodder) - methi - s' cane - ratoon - sunflower

Crop diversification in Haryana

Haryana with 4.4 m ha geographical area has 3.6 and 5.6 m ha net and gross cropped area respectively with about 155 per cent cropping intensity. During last 25 years food grain production has gone up from 25.92 lakh tonnes in 1966 to 102 lakh tonnes (about 4 times) during 1992-93 Similarly , production of cotton and oilseeds also gone up manifold. Area under different crops, cropping systems, shift in cropping systems, problems due to intensive use of inputs of monoculture, need for diversification and some suggested cropping systems of various farming situations of Haryana are discussed below.

1. Existing cropping systems

Rice wheat is most important cropping system occupying about 22% of cultivated area of the state. This is a major system in district Ambala (78%), Yamunanagar (80%), Kurukshetra (90%), Kaithal (89%), Karnal (96%), Panipat (96%) Sonipat (65%), Rohtak (8%), Sirsa (14%) and Hisar (14%). Another important cropping system in state is cotton- wheat occupying about 15.4 percent area. It is major cropping system in Sirsa (83%), Hisar(64%), Jind (32%), Bhiwani (16%), Rohtak (14%) and Kaithal (5%). Sugarcane based cropping system are mainly concentrated in irrigated areas adjointing to sugar mills, Bajra gram is an important cropping system in Bhiwani (67%) and Mahandergarh(79%), Bajra Mustard occupies major area in Mahendergarh (9%), Rewari (46%), Gurgaon (41%) Bhiwani (17%), Hisar (22%) and Sirsa (3%) Bajra - wheat is mainly followed in Faridabad (30%) and Jind (22%). Fodder based cropping systems are generally grown all over state in irrigated conditions. In dryland conditions of the State major area is occupied by single cropping of bajra, gram, rapeseed and mustard and other dryland crops.

In state as a whole, maximum area is under rice - wheat (21.67%), followed by cotton - wheat (15.4%), bajra - mustard (8.64/%) bajra gram (5.5%) fodder based cropping systems (7.80%) bajra fallow (1.66) fallow mustard (6.75) and fallow gram (4.25%). Other important cropping systems are maize wheat, jowar-wheat, jowar- gram and jowar - mustard.

2. Area, production, productivity and deviation in area under crops in different years.

Maximum shift in area during Kharif was in favour of rice which increased by about 73 per cent during 1981-82 to 1992-93. Area under bajra decreased drastically by about 3% and 25% in two decades, respectively. During rabi season major gainer was wheat crop with about 32 and 25% increase in area in two decades. Area under gram decreased by 6 and 63 percent but area under rapeseed and mustard increased by 24 and 176% in two decades, respectively. Other important gainer was cotton, but in case of sugarcane and potato, area increased between 1971-72 to 1981-82 but decreased later on. Area under jowar, maize and some other crops decreases in both decades.

3. Need for diversification

Increase in irrigation, facilities, availability of higher yielding varieties, intensive use of fertilizers and agro chemicals and practicing improved agro techniques have revolutionized crop production in Haryana during last two decades. Area, production and productivity of some crops increased manifold and total agriculture production also increased by about four times which made the state surplus in the a many commodities. High productivity of wheat, rice and cotton made these crops high economical and farmers started growing these crops as commercial crops in Northern, Eastern and Western parts of the state and these crops replaced other crops and occupy about 80 and 90 percent of cultivated area in some districts.

This shift in area, though helped on tremendous increase in agricultural production in state, resulted into specialized cropping and monoculture of some cropping system in major fertile and productive areas of the state. Of late, some problems have started surfacing in these areas. The problems are:

1. Decline in fertility and productivity of soil due to -
   i. Compaction of soil
   ii. Low organic matter content
   iii. Deficiency of macro and micro nutrients.
2. Problems of low and high water table.
3. Problem of soil as salinity
4. Shift in weed flora.
5. Resistance in some weeds to chemical weed control.
6. Problems of insect, pest and diseases.
7. Problem of pollution of soil, water air and agro products.
8. Problem or unemployment and under employments :

Keeping in view the available land and water resources, decline in soil fertility over exploitation of ground water resources in eastern zone and rise in water table in western zone, national priorities and farmer needs, diversification in agriculture in required for sustainability of the system and prosperity of the state. This has become more important due to monoculture practiced by farmers particularly in rice wheat belt where this cropping system is being practiced continuously for two decades. Crop wise diversification is discussed here and some important economic crop rotations have been suggested for different farming situations of the state.

4. (i) Diversification in Kharif Crops

Rice

Rice has become an important Kharif in the eastern zone of the state. If improved agro techniques are adopted then the average yield of the crop can be increased up to 4000 kg /ha by the end of the century and by improvement in average yield. State can produce 24 lakh tonnes of rice from 6 lakh hectares area by 2000 A.D. Moreover, due to problems of weed flora is rice wheat cropping system, there is urgent need to shift area from rice to other more remunerative crops like soybean, sugarcane, forage crops and vegetable crops.

Cotton

Cotton is one important cash crops in the western zone of the state. There is great scope for increasing area under cotton in the new command areas of Bhiwani and Mahendergarh districts. The area under desi and American cotton can be increased from 100 to 400 thousand hectares by the end of the century (table 1) .Some of the presently occupied area by the less efficient crops like bajra and jowar can be diverted to cotton with increasing irrigation facilities.

Bajra

Bajra is main food and fodder crop in dry regions of the state and its cultivations will continue in this belt though with reduced area of about 500 thousand hectares during 2000 AD. For this productivity can be increased to about 15 q/ha with the adoption of improved techniques. The area to be diverted from bajra can be put under guar, castor, pulses, fruits, crops and medicinal plants in low rainfall areas. However, arhar, soybean, groundnut and cotton can be grown where good quality irrigation water is available.

Maize

There is great demand of maize in starch industry and feed industry. The area under Kharif maize as well as winter can be increased profitably if cultivation of hybrid/ composite varieties along with improved input and techniques is adopted. Intercropping of soybean, black gram and green gram in Kharif and field pea, rajmah, gram, mustard in winter will further increase the returns.

Jowar

Jowar mostly is grown as a fodder crop in Haryana. Presently area under this crop will be retained either under fodder crops or near to cities it will go under vegetables and fruit crops.

Sugarcane

In near future the number of sugar mills in the state is likely to increase. Therefore, there is need to increase the area under sugarcane from 1.36 lakh to 2 lakh hectares by the end of century (Table 1). This area can be diverted from rice - wheat cropping system, particularly in areas having the problem of *Phalaris minor* showing resistance to chemical weed control.

(ii) Diversification of rabi crops

Wheat

Wheat is the most important rabi crop in the state. This is important crop in rice wheat cropping system as well. It is expected that rice wheat belt will have more problems of soil fertility and productivity, weeds, insect pest and disease and thus of sustainability and stability of productivity per unit area per unit times. Farmers in Karnal, Kurukshetra, Ambala, Kaithal, and Hisar districts are facing the problems of resistance in *Phalaris minor* to chemical weed control, One of the important factors of this problem is continuous monoculture of paddy wheat crop rotation. Therefore, there is urgent need of diversification of crops in these areas. By increasing the productivity of wheat crop, area under this crop can be reduced to 1500 th ha.

(iii) Diversification in pulses and oilseeds cultivation

Pulses and oilseeds are very sensitive crops because they are more susceptible to sudden changes in the weather conditions, insect pest and diseases. The area and production under these crops can be increased by growing high yielding crops like soybean, arhar, gram, field peas, sunflower, groundnut rapeseed and mustard, castor and linseed etc. they can also be raised in the off seasons like spring , summer, autumn. There is a great scope of growing pulses and oil seeds as intercrops with wide spaced crops like fruit crops sugarcane, cotton, castor, bajra, maize, potato and arhar etc. By 2000 AD the production of pulses can be increased by increasing area to 10 lakh hectares and productivity to 12 q/ha. Similarly, the area, production, and productivity of oilseeds can be increased to 6 lakh hectares, 10 lakh tonnes and 15 q/ha respectively. Some area under rice can be diverted for soybean and wheat area may be diverted to rapeseed and mustard and sunflower crops.

(iv) Diversification in area under fodder crops

To make the white revolution a great success, growing of green fodder in the plenty round the year is most important. At present about 5 percent of the total cropped area in the state is under fodder crops. The area can be increased to 700 thousand hectares (12 per cent) by the turn of the century. The production of nutrious fodders can also be increased by increasing the productivity through better management practices. Some area from paddy wheat crop rotation may be shifted to fodder crops.

(v) Diversification in fruits and vegetables crops area

Fruits and vegetables are rich sources of vitamins and minerals. At present area under fruits and vegetables is 72.7 thousand hectares. Cultivation of fruits, vegetables, flowers, etc. is very remunerative if assured market is available. The demand for fruits, vegetables, flowers, etc. increasing every day in the urban areas and therefore, there is need to increase area under these crops particularly in the vicinity cities. The area under these crops can be diverted from other less remunerative crops .By 2000 AD the area under fruits, vegetables, flowers etc. may be increased to 200 thousand hectares. These crops can also be included in the existing cropping systems. In problems area of rice - wheat cropping system belt, some area under wheat can be diverted to potato, onion and other vegetable crops.

(vi) Other supporting systems

i) Cultivation of flowers and mushrooms which find ready market in big cities.

ii) Bee keeping can be started with agro forestry, horticultural and field crops which would provide lot of nector of honey.

iii) Commercial seed production of field crops, vegetables and fodder is economical and some area can be diverted to seed production.

iv) Medicinal and aromatic plants like isbogol, rosa grass mulhati , lemon grass and spices like coriander, cumin, fennal, etc. can be planned in some area where other crops are being grown at present.

(vii) Economics of important cropping system

During seventies and early eighties, boost in productivity per unit area per unit time of paddy and wheat crops made these crops and cropping systems based on these crops more remunerative than pulses or oilseed based cropping systems. However, increase in cost of production and stagnated yield level of the major crops and also decrease in yield due to some problems like *Phalaris minor* in wheat, insect pest and disease attack in rice and insect attack in cotton resulted in thinking about some comparatively more economic and remunerative cropping systems. Some cropping system based on the pulse, oilseed, sugarcane, potato, fodder crops and vegetable crops based cropping system are either comparable or give more return than rice- wheat and cotton-wheat cropping systems.

Suggested approaches

Before we think of change through diversification, it has to be clearly under stood that any technology being economically viable and technically feasible will not find favour with its end users if it is not being supported by favorable public policies. Presently the public policies favour more and more production of food grains consumed in bulk quantities i.e. wheat and rice the Govt. was fully justified in its action because this was a necessity. But he did not make mid course corrections and the result is that farmers face difficulty in marketing due to more production but still he produces more food grains in excess of demand because it is remunerative for the farmers. On the other hand there is no motivation for diversification of product mix. i.e. edible oils, pulses and seed spices milk and milk product. Therefore, the first and foremost step for diversification to succeed is that the support price and procurement system must support diversification. Considering the complexities of our agriculture and policy, the Govt. intervention has to stay at least in the immediate future. Not only that, there must be a system of incentives and disincentives to support diversification.

Link prices and procurement with quality in order to boost export and improve farmers' national economy and to do that we need to act as follows:

(i) Creation of exclusive production zones

Some of the quality traits are being influenced more by the environment than the genetics. Location specific production technologies are available and some varieties in few crops are also having specific adaptation. These traits need to be utilized for better production, which will also help in better conservation to natural resources. For example we can have exclusive production Zone for Basmati Rice, Durum Wheat, Barley for malt, Vegetable, Floriculture and Dairy products etc. These production zones will have common seed source, common production technologies a system of procurement, processing, grading and marketing.

(ii) Value addition

After identification and creation of exclusive production zones, we need to provide services and infrastructure for value addition. For example, vegetables, potato (potato chips and potato finger) corn (corn flakes) fruits (canning preservation/ dyring etc), milk (milk products) edible oils and pulses (Processing and by products) and fortified foods (cereals & pulses).

(iii) Custom hiring service units and diversification in farm operation equipments/ machines

Considering some of the inherent traits of our agriculture such as ownership rights, holding size, nature and dependence, resource poor farmers etc the introduction of an element of commercialization is not possible. But it can be achieved by creating custom hiring service units for timely and effective farm operations. Where mechanization has become way of agriculture. These units can have trained youths in the rural areas (self help groups) that are to be supported by financial and technical institutions. Why farm machinery cannot be utilized as a means of irrigation system, zero tillage, furrow irrigation bed system, (FIRBS), bed planter? There is a need to conduct more research for further extension in the use of these machines. Why do not to explore the use of FIRBS for direct seeding of Basmati Rice to save labour, water, nutrients weedicides and avoid quality losses by reducing lodging?

(iv) Subsidy/ incentives / for crop diversification/ product mix

Some of the traditional crops like oilseeds and pulses used to have elements of instability in production and productivity. But the recent research development has reduced substantially the element of instability. The production technology is technically as well as economically viable. The cultivation of pulses will certainly add to the sustainability of the production system. But for the adoption of these technologies and boosting farmers confidence there is a need of incentives. Support prices needs to the matching with the open market prices government interventions commitment for procurement. These crops need to be considered the angle of what a producer gets and the consumer pays in respect of cereal and pulses.

14

Role of Organic Farming in Farming Systems

The increasing awareness of the deleterious effects of indiscriminate use of chemical fertilizers and pesticides in agriculture has led to the adoption of organic farming as an alternative method for conventional farming world-wide. It is gaining gradual momentum across the world and the growing awareness of organic food which is emerging as an attractive source of rural income generation especially in integrated farming systems.

In north west India where rice-wheat, cotton-wheat and pearl millet-wheat cropping systems are most prevalent. But now, these cropping systems are not proving farmers' friendly due to higher water requirement, declining yield, susceptibility to insects and pests and adverse effect on soil health. Moreover, in the new millennium with the growing demand of agricultural commodities in national as well as international market due to ever increasing population and globalization, Indian farmers are forced to produce higher quantum of quality agricultural commodities at low cost from shrinking land and natural resources.

Mungbean - wheat cropping system requires less water and hence, can be grown in areas where water availability is limited in addition to beneficial effect of mungbean to supply N to the succeeding wheat crop which are favorable for organic farming. Wheat (*desi*) also known as tall wheat requires less water and is very responsive to low doses of nutrients. The dwarf wheat grown by farmers with the application of chemicals in the form of fertilizers, weedicides, insecticides and pesticides is not preferred by people and therefore, the demand of organic wheat is rising fast in national and international market. This cropping system can be taken successfully in farming systems for higher monetary returns.

There is need to shift a portion of cultivable land with organic farming as an alternate practice technique. Organic farming gives emphasis on sustainability, environmental safety and low cost production techniques over chemical farming. Keeping the issues of 21st century agriculture in view following aspects also shows importance of organic farming: Reduce the toxic load; keep chemicals out of the air, water, soil and our bodies; reduce if not eliminate off farm pollution;

protect future generations; build healthy soil; taste better and truer flavor; assist farmers of all sizes; promote biodiversity; celebrate the culture of agriculture etc.

Though, after the advent of green revolution high yielding dwarf varieties of wheat are common with farmers of India. Yet the importance of growing wheat (*desi*) is still there especially in Madhya Pradesh, Haryana, Punjab, Uttar Pradesh states . It is in view of its low cost of cultivation, low nutrient requirement, high nutritional value, high market price, low preference and adaptability to climate and soils in the area.

Organic farming in area/zone/state

- Madhya Pradesh is largest promoter of O.F.
- Uttarakhand is recognized as organic zone of India.
- North Eastern and hill states has good opportunity to O.F.
- It is possible to export large quantity of tea, sugar, Coffee, cotton, rice, oilseeds, tobacco and fruits in Europe USA and middle east.
- Ministry of Commerce, Government of India launched National Programme for Organic Farming (NPOF) in 2000 and identified 6 agencies as accreditation agencies.

In general India and particularly Haryana and other northern states are traditionally organic in nature and in present agriculture along with FYM farmers have started growing pulses and legume forages which help in improving soil health and crop productivity. Green manuring has also been started in areas where rice-wheat cropping system is being followed for many years.

The areas adjoining Delhi where farmers take high value crops which fetch better market price organic farming is common. Organic farming is practiced in crops like vegetables, wheat (*desi*), basmati rice, pulses etc.

In south-west climatic zone of Haryana where soils are sandy loam having low organic matter the mungbean – wheat (*desi*) cropping system is being practiced by the farmers. This cropping system is beneficial to the farmers as the nutrient requirement of wheat (*desi*) is low and mungbean supplies nitrogen to the succeeding wheat crop.

Description of the problem: It is a matter of pride that India has attained self-sufficiency in food grain production. The increase in agricultural production has come through the release of high yielding varieties, which are responsive to fertilizers and irrigation. Presently, the fertilizer application has become a essential part of our agriculture. In many crops, yield stagnation has been reported with the present level of use of fertilizers. Also, a part of applied fertilizers is actually used by crops and remaining part is lost.

Use of pesticides on large scale has also contributed towards increased crop yields. As the yield plateau has been achieved in most crops, the increased use of agro-chemicals to increase the yield further may be dangerous due to their adverse effect on environment, human and animal health. Even the use of pesticides has not provided solutions to all the pest problems. About 0.1 per cent of the applied pesticides reaches to the targeted pests, while the remaining amount act as source of pollution of air, water, soil and take part in metabolic activities of microbial populations and other soil fauna. In view of this, there is an urgent need to provide an effective alternative to agro-chemicals, which can ensure increased yields as well as sustain the eco-system. Organic farming can play a key role in resource management and sustainable development of agriculture.

In the last decade dangers of agro-chemical has been realized and a need is being felt for organic farming. Presently two schools of thought are prevailing in respect of use/no use of agro-chemicals. One group is for the promotion of organic materials in partial/total replacement of agro-chemicals. Another strong group is of the opinion that the recommended levels of agro-chemical are not dangerous and their use is essential for increased agricultural production. This group is of the opinion that an attempt to discourage the use of agro-chemicals would adversely affect our food production. This seems to be correct in context of immediate short term gains but in broader perspective of sustainability of agriculture and to minimize the hazards of agro-chemicals on eco-system and mankind, the use of agro-chemicals needs to be minimized.

Not going in to details of these different school of thoughts it is certain that use of agro-chemicals is to be minimized by use of appropriate organic and INM/IPM practices in such a way that our agriculture production system have the capacity to produce enough good quality food on sustainable basis for the future generation.

Scope of organic farming: Organic farming was practiced since the period of ancient civilization, but the term organic farming was introduced by Lord Northbourne in 1940. International Forum for Organic Agriculture Movement (IFOAM) defined the organic agriculture as under:

Organic agriculture is the production system that sustains the health of soil eco-system and people by relying on ecological process, biodiversity and natural cycles and adapted condition than the use of inputs with adverse effects (IFOAM, 2008).

Organic farming involves the use of any type of organic material to supply plant nutrients, control of pests and diseases and to improve soil health for better crop production. It includes the various practices such as green manuring, green leaf manuring, microbial inoculation/bio-fertilizers, use of earthworms and other soil flora/fauna, crop rotations, intercropping etc. for improvement of soil health.

The pest control practices may include the use of bio-agents, predators, natural enemies, trap cropping etc. It is not restricted to the practice of supplying nutrient requirement from manure/organic sources alone or even controlling pests and diseases by non-chemical method alone but, it is an integral use of all possible organic sources and practices with the sole aim of strengthening crop husbandry.

Studies have shown that in initial 1 to 2 years of organic farming, yields may be reduced to some extent, but in long run it can produce the crop yields at par to chemical farming. The yield reduction due to organic farming has not been reported in many crops like pulses, oilseeds, vegetable crops. Also, tall wheat and basmati rice are known for their quality traits and requiring low nitrogen may perform well in organic cultivation. Whereas, in cotton, cereals, sugarcane etc. nitrogen requirement is high so the integrated nutrient management needs to be followed. The organic farming aims at developing self sustainable practices for increased crop yields and returns on long term basis. For its adoption a great deal of planning and physical work is required on the part of farmers.

Due to globalization process, it has become imperative for the farmers to grow quality agricultural produce for improving the economy of the state/country. Demand for quality food is increasing and there is intense competition for such foods in the market. To satisfy this, there is need to develop suitable organic package of practices of major cropping system including horticultural crops grown in the state. As the demand of organic food is on the rise so we have to explore the potential to produce organic food to exploit this. Cost effective organic food production technology need to be developed and standardized for important cropping system of the state to stay in competition with the developed nations.

What is organic farming?

Organic farming system in India is not new and is being followed from ancient time. It is a method of farming system which primarily aimed at cultivating the land and raising crops in such a way, as to keep the soil alive and in good health by use of organic wastes (crop, animal and farm wastes, aquatic wastes) and other biological materials along with beneficial microbes (biofertilizers) to release nutrients to crops for increased sustainable production in an eco friendly pollution free environment.

As per the definition of the United States Department of Agriculture (USDA) study team on organic farming "organic farming is a system which avoids or largely excludes the use of synthetic inputs (such as fertilizers, pesticides, hormones, feed additives etc) and to the maximum extent feasible rely upon crop rotations, crop residues, animal manures, off-farm organic waste, mineral grade rock additives and biological system of nutrient mobilization and plant protection".

FAO suggested that "Organic agriculture is a unique production management system which promotes and enhances agro-ecosystem health, including biodiversity, biological cycles and soil biological activity, and this is accomplished by using on-farm agronomic, biological and mechanical methods in exclusion of all synthetic off-farm inputs".

Need of organic farming

With the increase in population our compulsion would be not only to stabilize agricultural production but to increase it further in sustainable manner. The scientists have realized that the 'Green Revolution' with high input use has reached a plateau and is now sustained with diminishing return of falling dividends. Thus, a natural balance needs to be maintained at all cost for existence of life and property. The obvious choice for that would be more relevant in the present era, when these agrochemicals which are produced from fossil fuel and are not renewable and are diminishing in availability. It may also cost heavily on our foreign exchange in future.

The key characteristics of organic farming include

- Protecting the long term fertility of soils by maintaining organic matter levels, encouraging soil biological activity, and careful mechanical intervention
- Providing crop nutrients indirectly using relatively insoluble nutrient sources which are made available to the plant by the action of soil micro-organisms

- Nitrogen self-sufficiency through the use of legumes and biological nitrogen fixation, as well as effective recycling of organic materials including crop residues and livestock manures
- Weed, disease and pest control relying primarily on crop rotations, natural predators, diversity, organic manuring, resistant varieties and limited (preferably minimal) thermal, biological and chemical intervention
- The extensive management of livestock, paying full regard to their evolutionary adaptations, behavioural needs and animal welfare issues with respect to nutrition, housing, health, breeding and rearing
- Careful attention to the impact of the farming system on the wider environment and the conservation of wildlife and natural habitats.

Concept of organic farming

Organic farming is very much native to this land. History of organic farming will have to refer India and China because the farmers of these two countries are farmers of 20th centuries and it is organic farming that sustained them. This concept of organic farming is based on following principles:

- Nature is the best role model for farming, since it does not use any inputs nor demand unreasonable quantities of water.
- The entire system is based on intimate understanding of nature's ways. The system does not believe in mining of the soil nutrients and do not degrade it in any way for today's needs.
- The soil in this system is a living entity
- The soil's living population of microbes and other organisms are significant contributors to its fertility on a sustained basis and must be protected and nurtured at all cost.
- The total environment of the soil, from soil structure to soil cover is more important.

Present status of Organic Farming in India

- India ranks 33rd in the world in terms of area under organic cultivation
- 88th position in the world in terms of ratio of agriculture land under organic crops to total farming area
- About 5,85,970 tonnes of organic products worth of Rs 301 million are being exported from India
- Sikkim and Uttarakhand - organic state in India

Major Certified states

States	Area (Thousand ha)
Madhya Pradesh	163
Maharashtra	114
Orissa	74
J & K	32
Rajasthan	24
Kerala	14

Major products produced in India by organic farming:

Type	Products
Cereals	Rice, Wheat
Spices	Cardamom, Black pepper, ginger, turmeric, vanilla, tamarind, clove, nutmeg, mace chilly
Pulses	Pigeon pea, Urdbean, Soybean
Fruits	Mango, Banana, Pineapple, Grape, passion fruit, Orange, Cashew nut, walnut
Vegetables	Okra, Brinjal, Garlic, Onion, Tomato, Potato
Oilseeds	Sesame, castor, sunflower , mustard
Others	Cotton, herbal extract, Tea, Coffee

Organic Agriculture in India

Since January 1994 "Sevagram Declaration" for promotion of organic agriculture in India, organic farming has grown many folds and number of initiatives at Government and Non-Government level has given it a firm direction. While National Programme on Organic Production (NPOP) defined its regulatory framework, the National Project on Organic Farming (NPOF) has defined the promotion strategy and provided necessary support for area expansion under certified organic farming.

Growing certified area

Before the implementation of NPOP during 2001 and introduction of accreditation process for certification agencies, there was no institutional arrangement for assessment of organically certified area. Initial estimates during 2003-04 suggested that approximately 42,000 ha of cultivated land were certified organic. By 2009 India had brought more than 9.2 million ha of land under certification. Out of this while cultivable land was approximately 1.2 million ha, remaining 8 million ha area was forest land for wild collection. Growing awareness, increasing market demand, increasing inclination of farmers to go organic and growing institutional support has resulted into phenomenal growth in total certified area during the last five years. As on March 2009, total area under organic certification process stood at 12.01 lakh ha.

Important features of Indian organic sector

With the phenomenal growth in area under organic management and growing demand for wild harvest products India has emerged as the single largest country with highest arable cultivated land under organic management. India has also achieved the status of single largest country in terms of total area under certified organic wild harvest collection.

With the production of more than 77,000 MT of organic cotton lint India had achieved the status of largest organic cotton grower in the world a year ago, with more than 50% of total world's organic cotton.

Growing organic food market

Although no systematic information is available on size of organic food market by as per the survey conducted by the International Competence Centre for Organic Agriculture (ICCOA) in top 8 metro cities of India (which comprise about 5.3 % of the households) the market potential for organic foods in 2006 in top 8 metros of the country is at Rs 562 crore taking into account current purchase patterns of consumer in modern retail format. The overall market potential is estimated to be around Rs.1452 crore.

Growing area

Emerging from 42,000 ha under certified organic farming during 2003-04, the organic agriculture has grown almost 29 fold during the last 5 years. By March 2010 India has brought more than 4.48 million ha area under organic certification process. Out of this cultivated area accounts for 1.08 million ha while remaining 3.4 million ha is wild forest harvest collection area. Year wise growth of cultivated area under organic management is shown in Table below:

As on March 2014, India has brought 4.72 million ha area under organic certification process, which includes 0.6 million ha of cultivated agricultural land and 4.12 million ha of wild harvest collection area in forests. Growth of area under organic farming during different years is presented in Fig. 1a, and 1b. During 2012-13, India exported 165262 MT of organic products belonging to 135 commodities valuing at US$ 312 million (approximately INR 1900 crore). Domestic market is also growing at an annual growth rate of 15-25%. As per the survey conducted by ICCOA, Bangalore, domestic market during the year 2012-13 was worth INR 600 crore.

For quality assurance, India has internationally acclaimed certification process in place for export, import and domestic markets. During 2008-09, India produced about 18.78 lakh tonnes of certified organic products. Of this, nearly 54,000 tonne food items of worth Rs 591 crores were exported. Currently, India exports about 86 products worth over 100 million dollars to the world certified organic market registering a growth of over 30 per cent.

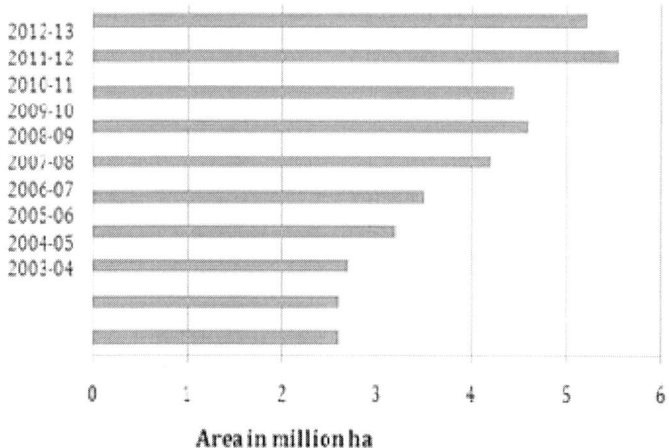

Fig. 1a : Total area under Certification (Cultivated + Wild harvest)

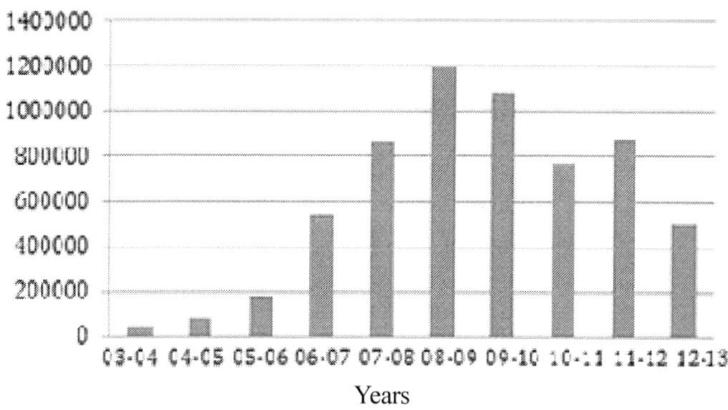

Fig. 1b : Cultivated area under organic certification (in ha)

Certified organic products including all varieties of food products namely basmati rice, pulses, honey, tea, spices, coffee, oil seeds, fruits, processed food, cereals, herbal medicines and their value added products are produced in India. Apart from edible sector, organic cotton fiber, garments, cosmetics, functional food products and body care products are also produced.

Banana, pomegranates, pineapple, grapes, amaranthus, ginger, large cardamom, sweet fennel, peanut, onion, sugar/ jaggery are other commodities which will emerge as significant organic commodities produced in India in the next few years.

According to FAO, organic farming would have long lasting, mostly beneficial effects on such important areas as:

Long-term productivity of the land: Protecting soils and enhancing their fertility would ensure productive capacity for future generations. Farmers often quote deteriorating soil quality as a major reason for adopting organic management. It can, therefore, be assumed that those farmers who adopted organic management practices found ways to improve the quality of their soil within the new management system, or at least stemmed the deterioration. Security of land tenure is important to the success of this task. If security is not guaranteed, there is little incentive for farmers to invest in a method that might only bring them income in the future rather than immediate rewards.

Food security and stability: In organic agriculture, a diversity of crops is often grown and many kinds of livestock kept. This diversification minimizes the risk of variation in production, as different crops react differently to climatic and edaphic variations, or have different times of growth (both in the time of the year and in length of the growth period). Consumers' demand for organic food and premium prices provide new export opportunities for farmers of the developing world, thus increasing their self-reliance. Organic agriculture can contribute to local food security in several ways. Organic farmers do not incur high initial expenses so less money is borrowed. Synthetic inputs, unaffordable to an increasing number of resource-poor farmers due to decreased subsidies and the need for foreign currency, are not used. Organic soil improvement may be the only economically sound system for resource poor, small-scale farmers.

Environmental impact: In a study with pesticides and fruit thinners used in apple production, showed that the total environmental impact rating of the conventional system was 6.2 times that of the organic one. Organic farmers forego the use of synthetic fertilizers. Most certification programs also restrict the use of mineral fertilizers, which can only be used to the extent necessary to supplement organic matter produced on the farm. There are environmental advantages to this: non-renewable fossil energy needs and N leaching is often reduced. Instead, farmers enhance soil fertility through use of manure (although the kind and its handling have significant effects on N content, and poor usage can create leaching problems), crop residues (e.g. corn stover, rice straw), legumes and green manures, and other natural fertilizers (e.g., rock phosphate, seaweed, guano, wood ash). Within the agricultural sector, dairy production systems represent the largest source of CH_4 and N_2O emissions and may therefore have a large potential for greenhouse gas mitigation. A study point out that SOM can significantly mitigate the greenhouse effect (e.g., via carbon sequestration). Disadvantages for not using synthetic chemicals must be considered as well: energy needs may escalate if thermal and mechanical weeding or intensive soil tillage is used. Many resource-poor farmers do not have access to livestock manure, often an important fertility component. Sometimes immature composts are used, which may contain pathogens and other contaminants. Finally, some areas in tropical countries may have such low soil fertility that synthetic inputs are essential. Soil protection techniques used in organic agriculture (e.g., terracing in the humid tropics, cover crops) combat soil erosion, compaction, salinization, and degradation of soils, especially through the use of crop rotations and organic materials that improve soil fertility and structure (including beneficial microbial influence and soil particle aggregation). Integrating trees and shrubs into the farming system also conserves soil and water and provides a defense against unfavorable weather conditions such as winds, droughts, and floods. Techniques used in organic agriculture also reduce water pollution and help conserve water on the farm. Although the benefits (both real and perceived) of organic farming and organic food are many, potential negative effects should also be noted, including the risk of contamination for human consumption. For example, nitrate leaching may contaminate ground water used for drinking, or organic livestock might be contaminated with disease-causing microorganisms from manure and by animal parasites.

Social impact: The social impact of organic farming is considerable as mentioned in the IFOAM's Principal Aims. The main benefit according by some organic farmers in developing countries (e.g., China and India) is that they now have better standards of living. Good product prices, low unemployment, dropped rural emigration, and reduced health risks (from chemicals) are the results of farming organic. In summary, the organic food movement apparently had its roots in a philosophy of life, beginning perhaps with Rudolf Steiner, a notable German thinker, in the 1920s. One of its common believes is that natural products are good, whereas man-made chemicals are not, or at least not as good as natural ones. This partially explains why organic farming avoids the use of synthetic fertilizers and pesticides. Certainly, organic farming has many benefits ranging from reduced environmental pollution to increased soil quality. Let us hope that organic farming will lead all farmers, and their consumers, toward a more productive, prosperous, sustainable, and healthy future.

Research findings

Currently, India ranks 10th among the top ten countries in terms of cultivable land under organic certification. The certified area includes 10% cultivable area with 0.50 mha and rest 90% (4.71 mha) is forest and wild area for collection of minor forest produces. The total area under organic certification is 5.21 mha (Anonymous, 2012-13). India produced around 1.34 mt of certified organic products which includes all varieties of food products namely Sugarcane, Cotton,

Basmati rice, Pulses, Tea, Spices, Coffee, Oil Seeds, Fruits and their value added products. Among all the states, Madhya Pradesh has covered largest area under organic certification followed by Rajasthan and Uttar Pradesh. India exported 135 products last year (Anonymous, 2012-13) with the total volume of 165262 mt including 4985 mt organic textiles. The organic agri export realization was around 374 million US $ including 160 US $ organic textiles registering a 4.38% growth over the previous year. At present there are 16 accredited inspection and certification agencies (Anonymous, 2012-13).

Presently, the natural resources are showing sign of fatigueless and it is the right time to focus all efforts to restore the soil, water, vegetation and environment status as prevailed earlier in the areas, where green revolution caused over-exploitation of resources and to increase per unit productivity by rationalizing the use of organic inputs. Likewise, in areas particularly in north-east part where the use of fertilizers and pesticides is already very less can be exploited for organic food production systems. It has also been highlighted that organic farming movement in India suffers from a lack of adequate institutional support in the areas of research , extension, certification and marketing, it require more scientific support than chemical farming. Increasing awareness about conservation of environment as well as health hazards associated with agro-chemicals and consumers‘ preferences, to safe and hazard free food are the major factors that lead to the growing interest in alternate forms of agriculture in the world. Organic farming is one of the broad spectrums of production systems, being considered in support of the environment and human health. The demand for organic food is steadily increasing both in the developed and developing countries with an annual average growth rate of 20-25% (NPOF, 2011-12). Organic farming is one of the fastest- growing sectors of agricultural production. The indiscriminate use of chemicals in crop production has deteriorated the soil health and polluted earth's environment. To reverse the situation, there is need to switch over to organic farming. Lack of standard package of practices, costly and complex certification process and inadequate marketing facilities had been identified as major constraints faced by the organic farmers.

Jat and Ahlawat (2010) concluded that application of FYM at 5.0 t/ha significantly increased the yield and yield attributes, nutrient (NPK) uptake in pigeonpea and groundnut, system productivity (1.71 t/ha), net returns (Rs 18,287) and available S in soil after harvest (15.72 kg/ha). However, B: C ratio was higher with crop receiving no FYM. Intercropping failed to influence the yield attributes, yield and nutrient uptake in pigeonpea, however, system productivity, net income and B: C ratio was higher in pigeonpea + groundnut system. Davari and Sharma (2010) find that, combinations of FYM + WR + B and VC + WR + B resulted in highest increase growth and yield attributing characters of rice and increased grain yield of basmati rice over control by 51-58%, and net return by 37-47%. These combinations were significantly superior to all other combinations in all the growth and yield parameters, yield, net profit and grain quality of basmati rice and thus hold a great promise in organic farming of basmati rice.

Continuous use of FYM, wheat straw and green manure in conjunction with fertilizers increased the soil organic carbon (OC), hydraulic conductivity, available N, P, K status and wheat yield, and decreased soil bulk density, soluble salts and pH (Kachroo *et. al*. 2006). Among organic materials, FYM resulted in highest OC (0.54%), available N (242.8 kg/ha), available P (17.7 kg/ha) and available K (318.5 kg/ha) level in soil. In general, the application of organic materials reduced the soil bulk density by 3.3% as compared to the fertilizers alone. Among different treatments, the grain and straw yield of wheat was highest where FYM (50% of N) was applied with fertilizers than the application of wheat straw, green manure or inorganic fertilizers alone. The results suggest that integrated use of inorganic fertilizer in combination with organic materials facilitates soil physical and chemical environment congenial for achieving higher crop productivity under pearl millet-wheat cropping system (Kumar *et. al*. 2012).

Studies from CCS HAU, Hisar in mungbean and wheat cropping system for nine years shown that the yield through application of organic sources ranged between 3455 to 5210kg ha^{-1}. Whereas integration of chemical and organic sources produced yield of 6282 kg ha^{-1}.The minimum decrease of 15.13 per cent in the system yield was recorded with application of 50 per cent recommended NPK + 50 %N through FYM as compared to recommended fertilizers. Net income was higher with the application of chemical fertilizers followed by integration of chemical and organic sources of nutrient as compared to organic sources of nutrients.

Literature Cited

Anonymous, (2012-13). Agricultural & Processed Food Products Export Development Authority (NPOP Annual Report), New Delhi - 110 016, India.

Baker C.J., Saxton K.E., & Ritchie W.R. (2002). *No-tillage seeding: science and practice*. 2nd Edition. CAB International. Oxford, UK.

Balusamy M., Shanmugham P.M., Baskaran R. (2003). Mixed farming an ideal farming. *Intensive Agric*. 41 (11–12): 20–25.

Davari, M.R. and Sharma, S.N. (2010). Effect of different combinations of organic materials and biofertilizers on productivity, grain quality and economics in organic farming of basmati rice (*Oryza sativa*). *Indian Journal of Agronomy* 55(4): 290 –294.

Gill M.S., Samra J.S., Singh Gurbachan, (2005). Integrated farming system for realizing high productivity under shallow water-table conditions. Research bulletins, Department of Agronomy, PAU, Ludhiana, pp. 1-29.

Gill M.S., Singh J.P., Gangwar K.S. (2009). Integrated farming system and agriculture sustainability. *Indian J Agron*. 54(2): 128-139.

Govindan, R., Chinnaswami, K.N. and Prince, J. (1990). Poultry-cum-fish culture in rice farming. *Indian J.Agron*. 35(1&2): 23-29.

Jat, R.A. and Ahlawat, I.P.S. (2010). Effect of organic manure and sulphur fertilization in pigeonpea (*Cajanus cajan*) + groundnut (*Arachis hypogaea*) intercropping system. *Indian Journal of Agronomy* 55(4): 276 –281.

Jayanthi C., Rangasamy A., Mythili S., Balusamy M., Chinnusamy C., Sankaran N. (2001). Sustainable productivity and profitability to integrated farming systems in low land farms. In: Extended summaries, pp. 79-81. (Eds: A.K. Singh, B. Gangwar, Pankaj and P.S. Pandey), National Symposium on Farming System Research on New Millennium, PDCSR, Modipuram.

Kachroo, D., Dixit, A.K. and Bali, A.S. (2006). Influence of crop residue, flyash and varying starter doses on growth, yield and soil characteristics in rice-wheat cropping system under irrigated conditions of Jammu region. *Indian Journal of Agricultural Sciences* 76: 3 6.

Kayombo, B. and R. Lal (1994). Responses of Tropical Crops to Soil Compaction; in: B.D. Soane and C. van Ouwerkerk (Eds.): Soil Compaction in Crop Production, Amsterdam 1994.

Korikanthimath V.S., Manjunath B.L. (2009). Integrated farming systems for sustainability in agricultural production. *Indian J Agron*. 54(2): 140-148.

Kumar, S., Dahiya, R., Kumar, P., Jhorar, B.S. and Phogat, V.K. (2012). Long - term effect of organic materials and fertilizers on soil properties in pearl millet-wheat cropping system. *Indian Journal of Agricultural Research*., 46(2): 161-166.

Lasson de A., Dolberg F. (1995). The casual effect of landholdings on livestock holdings. *Quarterly J IntAgr Germany*, Frankfurt. 24(4): 339-354.

Manjunath BL. (2002). Integrated farming systems for coastal region of Goa.Ph.D. Thesis Department of Agronomy, University of Agricultural Sciences, Dharwad.

Menegay, M.R., Hubbel, J.N. and William, R.D. (1978). Crop intensity index: A research method for measuring land use in multiple cropping. *Horticultural Science*, 61: 440-442.

NPOF, (2011-12). Annual Report of the National Project on Organic Farming. National Centre of Organic Fanning, Ghaziabad (UP).

Rangaswamy, A., Venkataswamy, R. and Palanniappan, S.P. (1996). Rice-poultry-fish-mushroom integrated farming systems for Tamilnadu. *Indian Journal of Agronomy*. 41(3): 344-348.

Singh C.B., Renkema J.A., Dhaka J.P., Singh, Keran, Schiere, J.B. (1993). Income and employment on small farmers.In : Proceeding An International workshop on Feeding of Ruminants on fibrous crop residues: Aspects of treatment, feeding, nutrient evaluation, research and extension. Karnal, Haryana, 4-8 February, 1991, pp. 67–76.

Singh Rajender, Singh Narinder, Phogat S.B., Sharma U.K., Singh R., Singh N. (1999). Income and employment potential of different farming system. *Haryana Agri Univ J Res*. 29 (3-4): 143–145.

Sivasankaran, D., Venkitiswamy, R. and Shanmugasundaram, V.S. (1995). A sustainable integrated fmg. Systems. *Madras Agric. J.* 82: 458-460.

Strought, A.M. (1975). Some definitional problems with multiple cropping diversifications. *Philippine Economist*, 14: 308-316.

Varughese K., Mathew T. (2009). Integrated farming systems for sustainability in coastal ecosystem. *Indian J Agron*. 54(2): 120-127.